The Gift

'The teaching of Marcel Mauss was one to which few can be compared. No acknowledgment of him can be proportionate to our debt.'

Claude Lévi-Strauss

'Marcel Mauss's famous *Essay on the Gift* becomes his own gift to the ages. Apparently completely lucid, with no secrets even for the novice, it remains a source of an unending ponderation . . .'

Marshall Sahlins, Stone Age Economics

'One could go so far as to say that a work as monumental as Marcel Mauss's *The Gift* speaks of everything but the gift: It deals with economy, exchange, contract (*do et des*), it speaks of raising the stakes, sacrifice, gift *and* countergift – in short, everything that in the thing itself impels the gift *and* the annulment of the gift.'

Jacques Derrida, Given Time

'*The Gift* is quite undeniably the masterwork of Marcel Mauss, his most justly famous writing, and the work whose influence has been the deepest.'

Claude Lévi-Strauss

Routledge Classics contains the very best of Routledge publishing over the past century or so, books that have, by popular consent, become established as classics in their field. Drawing on a fantastic heritage of innovative writing published by Routledge and its associated imprints, this series makes available in attractive, affordable form some of the most important works of modern times.

For a complete list of titles visit
www.routledgeclassics.com

Marcel
Mauss

The Gift

The form and reason for exchange
in archaic societies

With a foreword by Mary Douglas

 London and New York

Essai sur le don first published 1950 by Presses
Universitaires de France in *Sociologie et
Anthropologie*

English edition first published 1954
by Cohen & West
This translation first published 1990 by Routledge

First published in Routledge Classics 2002
by Routledge
2 Park Square, Milton Park, Abingdon, Oxon OX14 4RN

Reprinted 2004, 2005, 2006 (twice), 2007 (twice), 2008 (twice), 2009,
2010

Routledge is an imprint of the Taylor & Francis Group, an informa business

Translation © 1990 W. D. Halls
Foreword © 1990 Mary Douglas

Typeset in Joanna by RefineCatch Limited, Bungay, Suffolk
Printed and bound in Great Britain by
TJ International Ltd, Padstow, Cornwall

British Library Cataloguing in Publication Data
A catalogue record for this book is available from the British Library

ISBN10: 0–415–26748–X (hbk)
ISBN10: 0–415–26749–8 (pbk)

ISBN13: 978-0-415-26748-9 (hbk)
ISBN13: 978-0-415-26749-6 (pbk)

CONTENTS

EDITORIAL NOTE

The North American Indian term 'potlatch' has been retained in the translation. Various definitions of it are given in the text: 'system for the exchange of gifts', (as a verb) 'to feed, to consume', 'place of being satiated' [Boas]. As elaborated by Mauss, it consists of a festival where goods and services of all kinds are exchanged. Gifts are made and reciprocated with interest. There is a dominant idea of rivalry and competition between the tribe or tribes assembled for the festival, coupled occasionally with conspicuous consumption.

The French terms 'prestations' and 'contre-prestations' have no direct English equivalents. They represent, in the context in which they are used by Mauss, respectively the actual act of exchange of gifts and rendering of services, and the reciprocating or return of these gifts and services. Normally they have been referred to in the translation for brevity's sake, as 'total services' and 'total counter-services'.

It has not proved possible to reinstate the original English of the 170 quotations from English-language works, or presumed as such, used by Mauss. These works are from British, American, and Commonwealth sources and are often unidentifiable from the references given in the footnotes.

No free gifts

Mary Douglas

Charity is meant to be a free gift, a voluntary, unrequited surrender of resources. Though we laud charity as a Christian virtue we know that it wounds. I worked for some years in a charitable foundation that annually was required to give away large sums as the condition of tax exemption. Newcomers to the office quickly learnt that the recipient does not like the giver, however cheerful he be. This book explains the lack of gratitude by saying that the foundations should not confuse their donations with gifts. It is not merely that there are no free gifts in a particular place, Melanesia or Chicago for instance; it is that the whole idea of a free gift is based on a misunderstanding. There should not be any free gifts. What is wrong with the so-called free gift is the donor's intention to be exempt from return gifts coming from the recipient. Refusing requital puts the act of giving outside any mutual ties. Once given, the free gift entails no further claims from the recipient. The public is not deceived by free gift vouchers. For all the ongoing commitment the free-gift gesture has created, it might just as well never have happened. According to Marcel Mauss that is what is wrong with the free gift. A gift that does nothing to enhance solidarity is a contradiction.

Mauss says as much in reply to Bronislaw Malinowski who was surprised to find such precisely calculated return gifts in Melanesia. He evidently took with him to his fieldwork the idea that commerce and gift are two separate kinds of activity, the first based on exact recompense, the second spontaneous, pure of ulterior motive. Because the valuable things that circulated in the Trobriand Islands and a vast surrounding region were not in commercial exchange, he expected the transfers to fall into the category of gifts in his own culture. So he expended a lot of care in classifying gifts by the purity of the motives of the giver and concluded that practically nothing was given freely in this sense, only the small gift that a Trobriand husband regularly gave his wife could count. 'Pure gift? Nonsense!' declares Mauss: the Trobriand husband is actually recompensing his wife for sexual services. He would have said 'Nonsense!' just as heartily to Titmus's idea that the archetypal pure-gift relationship is the anonymous gift of blood,[1] as if there could be an anonymous relationship. Even the idea of a pure gift is a contradiction. By ignoring the universal custom of compulsory gifts we make our own record incomprehensible to ourselves: right across the globe and as far back as we can go in the history of human civilization, the major transfer of goods has been by cycles of obligatory returns of gifts.

Though this insight was taken up by archeologists and historians for reinterpreting antique systems of tax, revenues, and trade[2], a fancy archeological insight was not Mauss's objective. The *Essay on the Gift* was a part of an organized onslaught on contemporary political theory, a plank in the platform against utilitarianism. This intention is fully recognized in the new journal, *MAUSS*.[3] Mauss himself wrote very little about political philosophy but *The Gift* does not spring from nowhere; references to Emile Durkheim make quite clear where to look for the rest of the programme. And nor does Durkheim come from nowhere. First, I will explain the plan of the book, then I will place it in its context. Finally, I will indicate some of the work that has stemmed from it, and suggest what is still to be done to implement the original programme.

In this book the author has produced an idea that he has probably been mulling over for a long time. Indeed, the idea is profoundly original. We have seen how it runs against our established idea of gift. The book starts with describing the North American potlatch as an extreme form of an institution that is found in every region of the world. The potlatch is an example of a total system of giving. Read this too fast and you miss the meaning. Spelt out it means that each gift is part of a system of reciprocity in which the honour of giver and recipient are engaged. It is a total system in that every item of status or of spiritual or material possession is implicated for everyone in the whole community. The system is quite simple; just the rule that every gift has to be returned in some specified way sets up a perpetual cycle of exchanges within and between generations. In some cases the specified return is of equal value, producing a stable system of statuses; in others it must exceed the value of the earlier gift, producing an escalating contest for honour. The whole society can be described by the catalogue of transfers that map all the obligations between its members. The cycling gift system is the society.

The Gift is a grand exercise in positivist research, combining ethnology, history, and sociology. First Mauss presents the system as found in working order. This takes him to the ethnography of North America. What is striking about the potlatch among the Haïda and Tlingit of the Northwest coast is the extreme rivalry expressed by the rule always to return more than was received; failure to return means losing the competition for honour. There comes a point when there are just not enough valuable things to express the highest degrees of honour, so

conspicuous consumption is succeeded by conspicuous destruction. Then he turns to Melanesia where, in a less extreme form,
there are the essentials of potlatch, that is, totalized competitive
giving that incorporates in its cycles all things and services and
all persons. He treats Polynesia as a variant, because there the
totalized giving does not presume rivalry between donor and
recipient. When the paths of Polynesian gifts are traced, a stable,
hierarchical structure is revealed. It is not the competitive potlatch, but it is still a total system of gift. Where does the system
get its energy? In each case from individuals who are due to lose
from default heaping obloquy on defaulters and from beliefs
that the spirits would punish them. The system would not be
total if it did not include personal emotions and religion.

After presenting the system of gift functioning among American Indians and in Oceania, and among Eskimo and Australian
hunters, Mauss then turns to records of ancient legal systems.
Roman, Germanic, and other Indo-European laws all show signs
of the basic principles. There are no free gifts; gift cycles engage
persons in permanent commitments that articulate the dominant
institutions. Only after the full tour of ethnographic and legal
evidence do we finally reach the chapter on the theory of the gift
in classical Hindu law. Now we have definitely moved away from
working social systems to myths, legends, and fragments of laws:
not the system of gift but, as the chapter heading says, the theory
of gift. Mauss's early book with Henri Hubert (1889) on Sacrifice[4] took for its central theme a Vedic principle that sacrifice is a
gift that compels the deity to make a return: *Do ut des*; I give so
that you may give. Given the centrality of India in Max Muller's
philological speculations on mythology, any book at that time
on religion would need to study Hindu law and epic deeply. It
strikes me as likely that Mauss did get the idea of a morally
sanctioned gift cycle upholding the social cycle from the Vedic
literature that he studied in that first major research. I am
inclined to think that he harboured and developed the great idea

all those years. Certainly there is a close connection of matter and treatment between the two books.

In some histories of anthropology the main difference between old-fashioned folklore and modern ethnography has been identified as the replacement of library research by field-work. But I would suggest that the main important change came from a new criterion of sound analysis. The Gift was like an injunction to record the entire credit structure of a community. What a change that involved from current ideas about how to do ethnology can be seen by reading any of the earlier books cited in the voluminous footnotes whose unsystematic accounts of beliefs and ceremonies provided the uninterpreted bare bones of the gift system.

Because it starts from Northwest Coast American Indians and Melanesians and goes on to Polynesia and then to ancient texts, the book would seem to spring from the fusty debates of library researchers on comparative religion. Yet it is not about religion. It is about politics and economics. After the survey of evidence come the political and moral implications. Following Durkheim, Mauss also considered that every serious philosophical work should bear on public policy. The theory of the gift is a theory of human solidarity. Consequently, a brief reference to contemporary debates on health and unemployment insurance is in place, with the argument deduced from the preceding pages that the wage does not cover society's obligation to the worker. No obligations are ever completely covered. Though Mauss here refers approvingly to some English proposals on social policy, he is writing in a tradition strongly opposed to English liberal thought. At this point the Durkheimian context needs to be filled in.

The main strands in Durkheim's opposition to the English Utilitarians were already formulated by French political philosophers.[5] As Larry Siedentrop summarizes a tradition that stemmed from the eighteenth century, from Rousseau and

Tocqueville, it made three criticisms of English liberalism: first, that it was based on an impoverished concept of the person seen as an independent individual instead of as a social being; second, that it neglected how social relations change with changes in the mode of production; and third, that it had a too negative concept of liberty and so failed to appreciate the moral role of political participation. Furthermore, early English empiricist philosophy did not explain the role of social norms in shaping individual intentions and in making social action possible; their sensationalist model of the mind allowed no scope for explaining rule-governed action. Individualism is the essence of the French critique of utilitarianism. This is exactly where Durkheim's life work starts, as would appear from comparing his writings with the following paragraph by his biographer, Steven Lukes:[6]

> Benjamin Constant believed that 'when all are isolated by egoism, there is nothing but dust, and at the advent of a storm, nothing but mire',[7] while it was Alexis de Tocqueville who gave *individualisme* its most distinctive and influential liberal meaning in France. For Tocqueville it meant the apathetic withdrawal of individuals from public life into a private sphere and their isolation from one another, with a consequent and dangerous weakening of social bonds: individualism was
>
> > a deliberate and peaceful sentiment which disposes each citizen to isolate himself from the mass of his fellows . . . [which] at first saps only the virtues of public life, but, in the long run . . . attacks and destroys all others and is eventually absorbed into pure egoism.[8]
>
> (Lukes, 1973)

Among French socialists individualism was a bad word, referring to *laissez faire*, anarchy, social atomization, and exploitation of the poor under a regime of industrial capitalism. However, Durkheim's position was more complex. He believed that the success of a political system would depend on the extent to which it allowed individual self-awareness to flourish. He tried to keep a delicate balance between reproaching utilitarianism for overlooking that humans are social beings and reproaching socialism for overlooking the demands of the individual.

If one were to be forgetful of this traditional hostility to English utilitarianism it would be easy to misunderstand Durkheim's language and to fall into the trap of thinking that he really believed that society is a kind of separate intelligence that determines the thoughts and actions of its members as the mind does those of the body it is lodged in. Arguing against the nineteenth-century forms of utilitarianism, especially against the political philosophy of Herbert Spencer, it would have seemed hard for the anti-utilitarians to overestimate the importance of shared norms. And as for those whom he attacked, especially those across the Channel or across the Atlantic, it was evidently easier to misrepresent him than to disagree with what he was actually saying. Bartlett refers to Durkheim's idea of the collective memory as a quasimystic soul; Herbert Simon dissociates himself from Durkheimian 'group mind' implications; Alfred Schutz disdainfully dismisses Halbwachs' theories on the 'Collective Memory of Musicians' (which are very much the same as his own) because they are tainted by Durkheim's alleged theory of a unitary group consciousness; see also Bruno Latour on Durkheim's 'big animal'.[9] All these and many others forget that Durkheim's work was actually part of an ongoing research project with close collaborators who quite clearly did not give it this interpretation. So the counterattack has travestied versions of 'group mind', 'mystical unit', 'group psyche' that his language occasionally justifies but his precepts as to method certainly do not. This is why positivism was such an important plank in his programme. Positivism represented an attempt at objectivity. This is why it was necessary for Mauss to set out the plan of his book by beginning with the survey of functioning social systems, ending with Hindu texts about a vanished system (or one that had perhaps never existed in that form).

Today the same political debate is still engaged, between the contemporary utilitarians and those who, like Durkheim, deplore the effects of unfettered individualism. Some of those working in learned communities that embrace methodological individualism may be right to feel threatened by his teaching. Personally, I think it would be better for them to take it seriously. Hostility and a sense of threat are a sign that collective representations are at work. Our problem is how to take our own and other people's collective representations into account. Durkheim expected to do so by setting up sociology as a science, using positivist methods and looking for social facts. Science was to be a way of escaping bondage to past and to present loyalties. It is easy to mock his scientific pretensions, but who would deny that we really do need to seek for objectivity and to establish a responsible sociological discourse free of subjective hunches and concealed political pressure?

From this point of view The Gift rendered on extraordinary service to Durkheim's central project by producing a theory that could be validated by observation. For anthropologists the book has provided a basic requirement for modern fieldwork. It quickly became axiomatic that a field report would be below standard unless a complete account could be given of all transfers, that is, of all dues, gifts, fines, inheritances and successions, tributes, fees and payments; when this information is in place one also knows who gets left at the end of the day without honour or citizenship and who benefits from the cumulative transfers. With such a chart in hand the interpreter might be capable of sensing the meanings of ballads, calypsos, dirges, and litanies; without it one guess will do as well as any other.

Mauss rendered other inestimable services to Durkheim's project of a science of sociology. One is to have demonstrated that when the members of the Durkheimian school talked of society they did not mean an undecomposable unity, as many of their critics have supposed. If they had thought of society as an unanalysable, unchanging, sacralized entity, the researches of Durkheim's best pupils would never have been undertaken. *The Elementary Forms of the Religious Life*[10] gives snapshot pictures of Australian aborigines and American Indians worshipping spirits who sustain the social forms. It all seems very cut and dried. Durkheim and Mauss in *Primitive Classification*,[11] write as if categories are never negotiated but always come ready tailored to fit the institutions. Their argument at that point was not about change. They did in fact have a theory of change, that is, that changes in the organization of production radically transform the system of categories and beliefs.[12] If their theory had really been about a static social system, there would not have been any point in Maurice Halbwachs considering how public memory changes when part of the population goes away, taking its memories with it, or when a new influx comes bringing memories of their own past to the common pool.[13] Nor would Georges Davy have been so interested in the conditions under which oath-breaking is thought to be punished by God and those in which the sacredness of the oath diminishes.[14] It is an ignorant reading that supposes that Durkheim and his colleagues were looking for static correlations. The modern economy with its increasing specialization of functions is the backdrop to all these comparisons, and particularly to the gift system yielding place to the industrial system.

Another of Mauss's contributions to this collaborative effort is to have introduced a realistic idea of individuals in the pre-market social system where, according to Durkheim's formulations, one might expect only a community of humans mechanically connected to one another by their unquestioning use of this same ideas. Durkheim shared the common belief of his day in a gradual enriching and unfolding of the personality as the collective representations loosened their grip. However,

Mauss manages to incorporate individuals acting in their own interests, even in the kinds of societies in which Durkheim had thought that there was no scope for individual self-interest. On this Mauss rightly remarks that the concept of interest is itself modern.[15] He introduces psychology into the new sociology with essays on collective representations about death, about the body, and about the person.[16] In these he takes off from Durkheim's ideas and develops extended innovations upon them.

He also discovered a mechanism by which individual interests combine to make a social system, without engaging in market exchange. This is an enormous development beyond Durkheim's ideas of solidarity based on collective representations. The gift cycle echoes Adam Smith's invisible hand: gift complements market in so far as it operates where the latter is absent. Like the market it supplies each individual with personal incentives for collaborating in the pattern of exchanges. Gifts are given in a context of public drama, with nothing secret about them. In being more directly cued to public esteem, the distribution of honour, and the sanctions of religion, the gift economy is more visible than the market. Just by being visible, the resultant distribution of goods and services is more readily subject to public scrutiny and judgements of fairness than are the results of market exchange. In operating a gift system a people are more aware of what they are doing, as shown by the sacralization of their institutions of giving. Mauss's fertile idea was to present the gift cycle as a theoretical counterpart to the invisible hand. When anthropologists search around for a telling distinction between societies based on primitive and modern technologies, they try out various terms such as pre-literate, simple, traditional. Each has limitations that unfit it for general use. But increasingly we are finding that the idea of the gift economy comprises all the associations – symbolic, interpersonal, and economic – that we need for comparison with the market economy.

When I try to consider what would be needed now to

implement Mauss's original programme, I wonder which current ideas would be replaced if *The Gift* were to be as significant as he could have hoped. Where anthropology is concerned he would surely be more than satisfied. Nothing has been the same since. The big developments stem from this work. Before we had *The Gift*'s message unfolded for us we anthropologists, if we thought of the economy at all, treated it almost as a separate aspect of society, and kinship as separate again, and religion as a final chapter at the end. Evans-Pritchard, who promoted the original English translation and wrote a foreword to the edition that this one replaces, had Mauss's teaching very much at heart when he described the marriage dues of the Nuer as a strand in the total circulation of cattle, and wives, and children, and men: every single relationship had its substantiation in a gift.[17] This was a beginning, but there is no doubt that Claude Lévi-Strauss is the most indebted, which means of course that he gave counter-gifts as magnificent as he received. After *The Elementary Forms of Kinship*[18] we had to count transfers of men and women as the most important among the gifts in total symbolic systems. Numerous, very fine, comparative studies stand as testimony to the transformation of our outlook. However, it is not so easy to carry forward these analyses and apply them to ourselves.

The problem now is the same as it was for Mauss when it comes to applying his insights to contemporary, industrial society. Yet this is what he wanted to see done. As the last chapter in this volume shows, his own attempt to use the theory of the gift to underpin social democracy is very weak. Social security and health insurance are an expression of solidarity, to be sure, but so are a lot of other things, and there the likeness ends. Social democracy's redistributions are legislated for in elected bodies and the sums are drawn from tax revenues. They utterly lack any power mutually to obligate persons in a contest of honour. Taking the theory straight from its context in full-blown gift economies to a modern political issue was really jumping the gun. His own positivist method would require a great deal more patient spadework, both on theory and in collecting new kinds of data. I myself made an attempt to apply the theory of the gift to our consumption behaviour, arguing that it is much more about giving than the economists realize. Class structure would be clearly revealed in information about giving within and exclusion from reciprocal voluntary cycles of exchange. Much of the kind of information I needed about what happens in our society was missing from census and survey records.[19] It was information that could have been collected if Mauss's theory was recognized. If we persist in thinking that gifts ought to be free and pure, we will always fail to recognize our own grand cycles of exchanges, which categories get to be included and which get to be excluded from our hospitality.[20] More profound insights into the nature of solidarity and trust can be expected from applying the theory of the gift to ourselves. Though giving is the basis for huge industries, we cannot know whether it is the foundation of a circulating fund of stable esteem and trust, or of individualist competition as Thorstein Veblen thought.[21] We cannot know because the information is not collected in such a way as to relate to the issues.

I conclude by asking why this profound and original book had its impact mainly on small professional bodies of archeologists, classicists, and anthropologists. The answer might be that the debate with the utilitarians that Mauss was ready to enter before World War I had lost its excitement by the time he published this volume. One of the most fascinating topics in Lukes's biography is the relation of Durkheim's school to Marxism. Before the war the real enemy, the open enemy of French political philosophy, was Anglo-Saxon utilitarianism. After the war utilitarianism became the narrow province of a specialized discipline of economics. The political enemies of social democracy became communism and fascism. I have remarked how they traced a counterpoint to Marx's central ideas, neutralizing them as it were from communist taint and making something like Marxism safe for French democracy by diluting the revolutionary component.[22]

The political mood of the interwar years was dominated by concern for the erosion of civil liberties and excessive corporatist claims on the individual.

Now, however, the fashion has changed again. Utilitarianism is not just a technique of econometrics, nor a faded philosophy of the eighteenth century. Solidarity has again become a central topic in political philosophy. Social Darwinism walks again and the survival of the fittest is openly invoked. Philosophically creaking but technically shining, unified and powerful, utility theory is the main analytical tool for policy decisions. However, its intellectual assumptions are under attack. The French debate with the Anglo-Saxons can start again. This time round the sparks from Mauss's grand idea might well light a fuse to threaten methodological individualism and the idea of a free gift.

NOTES

1 R. Titmus (1970) *The Gift Relationship*, New York: Pantheon.
2 K. Polanyi, C.M. Arensberg, and H.W. Pearson (1957) *Trade and Market in Early Empires*, Glencoe: Free Press.
3 An acronym for Mouvement Anti-utilitariste dans les Sciences Sociales (New Series, vol. 1, 1988, La Découverte, Paris).
4 H. Hubert and M. Mauss (1899) 'Essai sur la nature et la fonction du sacrifice', *Année Sociologique*, 2: 29–138. (English translation by W.D. Halls, *Sacrifice: its Nature and Function*, with a Foreword by E.E. Evans-Pritchard, Routledge, London, 1964.)
5 L. Siedentrop (1979) 'Two Liberal Traditions' in A. Ryan (ed.) *The Idea of Freedom*, Oxford: Clarendon Press, pp. 153–74. Starting from Rousseau in the eighteenth century, and with Condillac, Bonald, and Maistre, Larry Siedentrop names as the nineteenth-century protagonists of this criticism Madame de Staël, Benjamin Constant, and Les Doctrinaires. The latter group included Guizot and de Tocqueville who

took the critique of political theory as an urgent post-revolutionary reform. There was more than a touch of political reaction in the movement. The Doctrinaire theorists were strongly committed to the idea of hierarchy and the Doctrinaire government (1815–20 and 1820–7) tried to restore the conditions of the *Ancien Régime*.

6 S. Lukes (1973) *Emile Durkheim, His Life and Work*, London: Allen Lane pp. 197–8.

7 B. Constant quoted in Lukes (1973), himself quoting H. Marion (n.d.). 'Individualisme', in *La Grande Encyclopédie*, vol. xx, Paris.

8 A. de Tocqueville (1835–40) *De la démocratie en Amérique*, ii, 2, Ch. 11 in *Oeuvres Complètes*, J.P. Mayer (ed.) (1951) Paris, t. 1, pt. 2:10.

9 Bartlett (1932) *Remembering: A Study in Experimental and Social Psychology*, Cambridge: Cambridge University Press.
   H. Simon (1945) *Administrative Behavior, A Study of Decision-making Processes in Administrative Organisation*, Glencoe: Free Press.
   A. Schutz (1951) 'Making Music Together' in *Collected Papers*, 1–3, The Hague: Martinus Nijhoff.
   B. Latour (1988) in review of Mary Douglas's *How Institutions Think*, *Contemporary Sociology, an International Journal of Reviews*: 383–5.

10 E. Durkheim (1912) *Les Formes Elémentaires de la Vie Religieuse*, Paris: Alcan.

11 E. Durkheim and M. Mauss (1903) 'De quelques formes primitives de classification: contribution à l'etude des representations collectives', *l'Année Sociologique* 6.

12 E. Durkheim (1893) *De la Division du Travail Social: étude sur l'organisation des sociétés superieures*, Paris: Alcan.

13 M. Halbwachs (1925) *Les Cadres Sociaux de la Mémoire*, Paris: Alcan.

14 G. Davy (1922) 'La foi jurée', *Étude Sociologique du Problème du Contrat, la Formation du lien contractuel*, Paris. (English translation by W.D. Halls, *The Division of Labour in Society*, Macmillan, London and New York, 1984.)

15 See A. Hirschman (1973) *The Passions and the Interests*, Princeton, N.J.: Princeton University Press:

16 M. Mauss (1926) 'L'idée de mort', *Journal de Psychologie Normale et Pathologique*,
   M. Mauss (1936) 'Les techniques du corps', *Journal de Psychologie*, 3–4.

17 E.E. Evans-Pritchard (1940) *The Nuer: The Political Institutions of a Nilotic People*, Oxford: Clarendon Press.

E.E. Evans-Pritchard (1951) *Kinship and Marriage among the Nuer*, Oxford: Clarendon Press.

18 C. Lévi-Strauss (1949) *Les Structures Elémentaires de la Parenté*, Paris: Presses Universitaires de France.

19 M. Douglas (1978) *The World of Goods*, London: Basic Books and Penguin.

20 M. Douglas (ed.) (1984) *Food in the Social Order*, New York: Russell Sage Foundation.

21 T. Veblen (1928) *The Theory of the Leisure Class*, New York: Vanguard Press.

22 M. Douglas (1980) 'Introduction: Maurice Halbwachs (1877–1945)', *The Collective Memory*, New York: Harper and Row.

INTRODUCTION

THE GIFT, AND ESPECIALLY THE OBLIGATION TO RETURN IT

Epigraph

Below we give a few stanzas from the Havamal, one of the old poems of the Scandinavian Edda.[1] They may serve as an epigraph for this study, so powerfully do they plunge the reader into the immediate atmosphere of ideas and facts in which our exposition will unfold.[2]

(39)   I have never found a man so generous
         And so liberal in feeding his guests
         That 'to receive would not be received',
         Nor a man so . . . [the adjective is missing]
         Of his goods
         That to receive in return was disagreeable to
         Him[3]

(41)   With weapons and clothes
       Friends must give pleasure to one another;
       Everyone knows that for himself [through his
       Own experience].
       Those who exchange presents with one another
       Remain friends the longest
       If things turn out successfully.

reciprocity a basis for friendship

(42)   One must be a friend
       To one's friend,
       And give present for present;
       One must have
       Laughter for laughter
       And sorrow for lies

(44)   You know, if you have a friend
       In whom you have confidence
       And if you wish to get good results
       Your soul must blend in with his
       And you must exchange presents
       And frequently pay him visits.

(44)   But if you have another person
(sic) Whom you mistrust
       And if you wish to get good results,
       You must speak fine words to him
       But your thoughts must be false
       And you must lament in lies.

(46)   This is the way with him
       In whom you have no trust
       And whose sentiments you suspect,
       You must smile at him
       And speak in spite of yourself:
       Presents given in return must be similar to
       Those received.

reciprocity even among enemies

all human interactions are based on exchange / reciprocity

(47)   Noble and valiant men
         Have the best life;
         They have no fear at all
         But a coward fears everything:
         The miser always fears presents.

Cahen also points out to us stanza 145:

(145) It is better not to beg [ask for something]
       Than to sacrifice too much [to the gods]:
       A present given always expects one in return.
       It is better not to bring any offering
       Than to spend too much on it.

Careful what you wish for

Programme

The subject is clear. In Scandinavian civilization, and in a good number of others, exchanges and contracts take place in the form of presents; in theory these are voluntary, in reality they are given and reciprocated obligatorily.

The present monograph is a fragment of more extensive studies. For years our attention has been concentrated on both the organization of contractual law and the system of total economic services operating between the various sections or subgroups that make up so-called primitive societies, as well as those we might characterize as archaic. This embraces an enormous complex of facts. These in themselves are very complicated. Everything intermingles in them, everything constituting the strictly social life of societies that have preceded our own, even those going back to protohistory. In these 'total' social phenomena, as we propose calling them, all kinds of institutions are given expression at one and the same time – religious, juridical, and moral, which relate to both politics and the family; likewise economic ones, which suppose special forms of production and consumption, or rather, of performing total services and of distribution. This is not to take into account the aesthetic phenomena to which these facts lead, and the contours of the phenomena that these institutions manifest.

Among all these very complex themes and this multiplicity of social 'things' that are in a state of flux, we seek here to study only one characteristic – one that goes deep but is isolated: the so to speak voluntary character of these total services, apparently free and disinterested but nevertheless constrained and self-interested. Almost always such services have taken the form of the gift, the present generously given even when, in the gesture accompanying the transaction, there is only a polite fiction, formalism, and social deceit, and when really there is obligation and economic self-interest. Although we shall indicate in detail all the various principles that have imposed this appearance on a necessary form of exchange, namely, the division of labour in society itself – among all these principles we shall nevertheless study only one in depth. *What rule of legality and self-interest, in societies of a backward or archaic type, compels the gift that has been received to be obligatorily reciprocated? What power resides in the object given that causes its recipient to pay it back?* This is the problem on which we shall fasten more particularly, whilst indicating others. By examining a fairly large body of facts we hope to respond to this very precise question and to point the way to how one may embark upon a study of related questions. We shall also see to what fresh problems we are led. Some concern a permanent form of contractual morality, namely, how the law relating to things even today remains linked to the law relating to persons. Others deal with the forms and ideas that, at least in part, have always presided over the act of exchange, and that even now partially complement the notion of individual self-interest.

We shall thus achieve a dual purpose. On the one hand, we shall arrive at conclusions of a somewhat archeological kind concerning the nature of human transaction in societies around us, or that have immediately preceded our own. We shall describe the phenomena of exchange and contract in those societies that are not, as has been claimed, devoid of economic markets – since the market is a human phenomenon that, in our view, is not foreign to any known society – but whose system of exchange is different from ours. In these societies we shall see the market as it existed before the institution of traders and before their main invention – money proper. We shall see how it functioned both before the discovery of forms of contract and sale that may be said to be modern (Semitic, Hellenic, Hellenistic, and Roman), and also before money, minted and inscribed. We shall see the morality and the organization that operate in such transactions.

As we shall note that this morality and organization still function in our own societies, in unchanging fashion and, so to speak, hidden, below the surface, and as we believe that in this we have found one of the human foundations on which our societies are built, we shall be able to deduce a few moral conclusions concerning certain problems posed by the crisis in our own law and economic organization. There we shall call a halt. This page of social history, of theoretical sociology, of conclusions in the field of morality, and of political and economic practice only leads us after all to pose once more, in different forms, questions that are old but ever-new.[4]

Method

We have followed the method of exact comparison. First, as always, we have studied our subject only in relation to specific selected areas: Polynesia, Melanesia, the American Northwest, and a few great legal systems. Next, we have naturally only chosen those systems of law in which we could gain access, through documents and philological studies, to the consciousness of the societies themselves, for here we are dealing in words and ideas. This again has restricted the scope of our comparisons. Finally, each study focused on systems that we have striven to describe each in turn and in its entirety. Thus we have renounced that continuous comparison in which everything is mixed up together, and in which institutions lose all local colour and documents their savour.[5]

THE RENDERING OF TOTAL SERVICES. THE GIFT AND POTLATCH

The present study forms part of a series of researches that Davy and myself have been pursuing for a long time, concerning the archaic forms of contract.[6] A summary of these is necessary.

Apparently there has never existed, either in an era fairly close in time to our own, or in societies that we lump together somewhat awkwardly as primitive or inferior, anything that might resemble what is called a 'natural' economy.[7] Through a strange but classic aberration, in order to characterize this type of economy, a choice was even made of the writings by Cook relating to exchange and barter among the Polynesians.[8] Now, it is these same Polynesians that we intend to study here. We shall see how far removed they are from a state of nature as regards law and economics.

In the economic and legal systems that have preceded our own, one hardly ever finds a simple exchange of goods, wealth, and products in transactions concluded by individuals. First, it is not individuals but collectivities that impose obligations of exchange and contract upon each other.[9] The contracting parties are legal entities: clans, tribes, and families who confront and oppose one another either in groups who meet face to face in one spot, or through their chiefs, or in both these ways at once.[10] Moreover, what they exchange is not solely property and wealth, movable and immovable goods, and things economically useful.

In particular, such exchanges are acts of politeness: banquets, rituals, military services, women, children, dances, festivals, and fairs, in which economic transaction is only one element, and in which the passing on of wealth is only one feature of a much more general and enduring contract. Finally, these total services and counter-services are committed to in a somewhat voluntary form by presents and gifts, although in the final analysis they are strictly compulsory, on pain of private or public warfare. We propose to call all this the *system of total services*. The purest type of such institutions seems to us to be characterized by the alliance of two phratries in Pacific or North American tribes in general, where rituals, marriages, inheritance of goods, legal ties and those of self-interest, the ranks of the military and priests – in short everything, is complementary and presumes co-operation between the two halves of the tribe. For example, their games, in particular, are regulated by both halves.[11] The Tlingit and the Haïda, two tribes of the American Northwest, express the nature of such practices forcefully by declaring that 'the two tribal phratries show respect to each other'.[12]

But within these two tribes of the American Northwest and throughout this region there appears what is certainly a type of these 'total services', rare but highly developed. We propose to call this form the 'potlatch', as moreover, do American authors using the Chinook term, which has become part of the everyday language of Whites and Indians from Vancouver to Alaska. The word potlatch essentially means 'to feed', 'to consume'.[13] These tribes, which are very rich, and live on the islands, or on the coast, or in the area between the Rocky Mountains and the coast, spend the winter in a continual festival of feasts, fairs, and markets, which also constitute the solemn assembly of the tribe. The tribe is organized by hierarchical confraternities and secret societies, the latter often being confused with the former, as with the clans. Everything – clans, marriages, initiations, Shamanist seances and meetings for the worship of the great gods, the totems or the collective or individual ancestors of the clan – is woven into an inextricable network of rites, of total legal and economic services, of assignment to political ranks in the society of men, in the tribe, and in the confederations of tribes, and even internationally.[14] Yet what is noteworthy about these tribes is the principle of rivalry and hostility that prevails in all these practices. They go as far as to fight and kill chiefs and nobles. Moreover, they even go as far as the purely sumptuary destruction of wealth that has been accumulated in order to outdo the rival chief as well as his associate (normally a grandfather, father-in-law, or son-in-law).[15] There is total service in the sense that it is indeed the whole clan that contracts on behalf of all, for all that it possesses and for all that it does, through the person of its chief.[16] But this act of 'service' on the part of the chief takes on an extremely marked agonistic character. It is essentially usurious and sumptuary. It is a struggle between nobles to establish a hierarchy amongst themselves from which their clan will benefit at a later date.

We propose to reserve the term potlatch for this kind of institution that, with less risk and more accuracy, but also at greater length, we might call: *total services of an agonistic type.*

Up to now we had scarcely found any examples of this institution except among the tribes of the American Northwest,[17] Melanesia, and Papua.[18] Everywhere else, in Africa, Polynesia, Malaysia, South America, and the rest of North America, the basis of exchanges between clans and families appeared to us to be the more elementary type of total services. However, more detailed research has now uncovered a quite considerable number of intermediate forms between those exchanges comprising very acute rivalry and the destruction of wealth, such as those of the American Northwest and Melanesia, and others, where emulation is more moderate but where those entering into contracts seek to outdo one another in their gifts. In the same way we vie with one another in our presents of thanks, banquets and weddings, and in simple invitations. We still feel the need to *revanchieren*,[19] as the Germans say. We have discovered intermediate forms in the ancient Indo-European world, and especially among the Thracians.[20]

Various themes – rules and ideas – are contained in this type of law and economy. The most important feature among these spiritual mechanisms is clearly one that obliges a person to reciprocate the present that has been received. Now, the moral and religious reason for this constraint is nowhere more apparent than in Polynesia. Let us study it in greater detail, and we will plainly see what force impels one to reciprocate the thing received, and generally to enter into real contracts.

1

THE EXCHANGE OF GIFTS AND THE OBLIGATION TO RECIPROCATE (POLYNESIA)

I
'TOTAL SERVICES', 'MATERNAL* GOODS' AGAINST 'MASCULINE GOODS'† (SAMOA)

During this research into the extension of contractual gifts, it seemed for a long time as if potlatch proper did not exist in Polynesia. Polynesian societies in which institutions were most comparable did not appear to go beyond the system of 'total services', permanent contracts between clans pooling their women, men, and children, and their rituals, etc. We then studied in Samoa the remarkable custom of exchanging emblazoned

* The French *utérin*, strictly speaking, relates to children of the same mother, but not necessarily of the same father. It is translated as 'maternal' and relates to the goods that are passed on to such children, i.e. 'maternal goods'.
† 'Masculine goods' [*biens masculins*] relates to goods passed on to children through the father's side.

matting between chiefs on the occasion of a marriage, which did not appear to us to go beyond this level.[1] The elements of rivalry, destruction, and combat appeared to be lacking, whereas this was not so in Melanesia. Finally, there were too few facts available. Now we would be less critical about the facts.

First, this system of contractual gifts in Samoa extends far beyond marriage. Such gifts accompany the following events: the birth of a child,[2] circumcision,[3] sickness,[4] a daughter's arrival at puberty,[5] funeral rites,[6] trade.[7]

Next, two essential elements in potlatch proper can be clearly distinguished here: the honour, prestige, and *mana* conferred by wealth;[8] and the absolute obligation to reciprocate these gifts under pain of losing that *mana*, that authority – the talisman and source of wealth that is authority itself.[9]

On the one hand, as Turner tells us:

> After the festivities at a birth, after having received and reciprocated the *oloa* and the *tonga* – in other words, masculine and feminine goods – husband and wife did not emerge any richer than before. But they had the satisfaction of having witnessed what they considered to be a great honour: the masses of property that had been assembled on the occasion of the birth of their son.[10]

On the other hand, these gifts can be obligatory and permanent, with no total counter-service in return except the legal status that entails them. Thus the child whom the sister, and consequently the brother-in-law, who is the maternal uncle, receive from their brother and brother-in-law to bring up, is himself termed a *tonga*, a possession on the mother's side.[11] Now, he is:

> the channel along which possessions that are internal in kind,[12] the *tonga*, continue to flow from the family of the child to that family. Furthermore, the child is the means whereby his parents

> can obtain possessions of a foreign kind (*oloa*) from the parents who have adopted him, and this occurs throughout the child's lifetime.

> This sacrifice [of the natural bonds] facilitates an easy system of exchange of property internal and external to the two kinship sides.

In short, the child, belonging to the mother's side, is the channel through which the goods of the maternal kin are exchanged against those of the paternal kin. It suffices to note that, living with his maternal uncle, the child has plainly the right to live there, and consequently possesses a general right over the latter's possessions. This system of 'fosterage' appears very close to that of the generally acknowledged right of the maternal nephew in Melanesian areas over the possessions of his uncle.[13] Only the theme of rivalry, combat, and destruction is lacking, for there to be potlatch.

Let us, however, note these two terms, *oloa*, and *tonga*, and let us consider particularly the *tonga*. This designates the permanent paraphernalia, particularly the mats given at marriage,[14] inherited by the daughters of that marriage, and the decorations and talismans that through the wife come into the newly founded family, with an obligation to return them.[15] In short, they are kinds of fixed property – immovable because of their destination. The *oloa*[16] – designate objects, mainly tools, that belong specifically to the husband. These are essentially movable goods. Thus nowadays this term is applied to things passed on by Whites.[17] This is clearly a recent extension of the meaning. We can leave on one side Turner's translation: *oloa* = foreign; *tonga* = native. It is incorrect and insufficient, but not without interest, since it demonstrates that certain goods that are termed *tonga* are more closely linked to the soil,[18] the clan, the family, and the person than certain others that are termed *oloa*.

Yet, if we extend the field of our observation, the notion of tonga immediately takes on another dimension. In Maori, Tahitian, Tongan, and Mangarevan (Gambier), it connotes everything that may properly be termed possessions, everything that makes one rich, powerful, and influential, and everything that can be exchanged, and used as an object for compensating others.[19] These are exclusively the precious articles, talismans, emblems, mats, and sacred idols, sometimes even the traditions, cults, and magic rituals. Here we link up with that notion of property-as-talisman, which we are sure is general throughout the Malaysian and Polynesian world, and even throughout the Pacific as a whole.[20]

II
THE SPIRIT OF THE THING GIVEN (MAORI)

This observation leads us to a very important realization: the taonga [sic] are strongly linked to the person, the clan, and the earth, at least in the theory of Maori law and religion. They are the vehicle for its mana, its magical, religious, and spiritual force. In a proverb that happily has been recorded by Sir George Grey[21] and C.O. Davis[22] the taonga are implored to destroy the individual who has accepted them. Thus they contain within them that force, in cases where the law, particularly the obligation to reciprocate, may fail to be observed.

Our much regretted friend Hertz had perceived the importance of these facts. With his touching disinterestedness he had noted down 'for Davy and Mauss', on the card recording the following fact. Colenso says:[23] 'They had a kind of exchange system, or rather one of giving presents that must ultimately either be reciprocated or given back.' For example, dried fish is exchanged for jellied birds or matting.[24] All these are exchanged between tribes or 'friendly families without any kind of stipulation'.

But Hertz had also noted – and I have found it among his records – a text whose importance had escaped the notice of both of us, for I was equally aware of it.

Concerning the *hau*, the spirit of things, and especially that of the forest and wild fowl it contains, Tamati Ranaipiri, one of the best Maori informants of Elsdon Best, gives us, completely by chance, and entirely without prejudice, the key to the problem.[25]

> I will speak to you about the *hau* ... The *hau* is not the wind that blows – not at all. Let us suppose that you possess a certain article (*taonga*) and that you give me this article. You give it me without setting a price on it.[26] We strike no bargain about it. Now, I give this article to a third person who, after a certain lapse of time, decides to give me something as payment in return (*utu*).[27] He makes a present to me of something (*taonga*). Now, this *taonga* that he gives me is the spirit (*hau*) of the *taonga* that I had received from you and that I had given to him. The *taonga* that I received for these *taonga* (which came from you) must be returned to you. It would not be fair (*tika*) on my part to keep these *taonga* for myself, whether they were desirable (*rawe*) or undesirable (*kino*). I must give them to you because they are a *hau*[28] of the *taonga* that you gave me. If I kept this other *taonga* for myself, serious harm might befall me, even death. This is the nature of the *hau*, the *hau* of personal property, the *hau* of the *taonga*, the *hau* of the forest. *Kati ena* (But enough on this subject).

This text, of capital importance, deserves a few comments. It is purely Maori, permeated by that, as yet, vague theological and juridical spirit of doctrines within the 'house of secrets', but at times astonishingly clear, and presenting only one obscure feature: the intervention of a third person. Yet, in order to understand fully this Maori juridical expert, one need only say:

The *taonga* and all goods termed strictly personal possess a *hau*, a spiritual power. You give me one of them, and I pass it on to a third party; he gives another to me in turn, because he is impelled to do so by the *hau* my present possesses. I, for my part, am obliged to give you that thing because I must return to you what is in reality the effect of the *hau* of your *taonga*.

When interpreted in this way the idea not only becomes clear, but emerges as one of the key ideas of Maori law. What imposes obligation in the present received and exchanged, is the fact that the thing received is not inactive. Even when it has been abandoned by the giver, it still possesses something of him. Through it the giver has a hold over the beneficiary just as, being its owner, through it he has a hold over the thief.[29] This is because the *taonga* is animated by the *hau* of its forest, its native heath and soil. It is truly 'native':[30] the *hau* follows after anyone possessing the thing.

It not only follows after the first recipient, and even, if the occasion arises, a third person, but after any individual to whom the *taonga* is merely passed on.[31] In reality, it is the *hau* that wishes to return to its birthplace, to the sanctuary of the forest and the clan, and to the owner. The *taonga* or its *hau* – which itself moreover possesses a kind of individuality[32] – is attached to this chain of users until these give back from their own property, their *taonga*, their goods, or from their labour or trading, by way of feasts, festivals and presents, the equivalent or something of even greater value. This in turn will give the donors authority and power over the first donor, who has become the last recipient. This is the key idea that in Samoa and New Zealand seems to dominate the obligatory circulation of wealth, tribute, and gifts.

Such a fact throws light upon two important systems of social phenomena in Polynesia and even outside that area. First, we can grasp the nature of the legal tie that arises through the passing on of a thing. We shall come back presently to this point, when

gifts are a part of the giver

we show how these facts can contribute to a general theory of obligation. For the time being, however, it is clear that in Maori law, the legal tie, a tie occurring through things, is one between souls, because the thing itself possesses a soul, is of the soul. Hence it follows that to make a gift of something to someone is to make a present of some part of oneself. Next, in this way we can better account for the very nature of exchange through gifts, of everything that we call 'total services', and among these, potlatch. In this system of ideas one clearly and logically realizes that one must give back to another person what is really part and parcel of his nature and substance, because to accept something from somebody is to accept some part of his spiritual essence, of his soul. To retain that thing would be dangerous and mortal, not only because it would be against law and morality, but also because that thing coming from the person not only morally, but physically and spiritually, that essence, that food,[33] those goods, whether movable or immovable, those women or those descendants, those rituals or those acts of communion – all exert a magical or religious hold over you. Finally, the thing given is not inactive. Invested with life, often possessing individuality, it seeks to return to what Hertz called its 'place of origin' or to produce, on behalf of the clan and the native soil from which it sprang, an equivalent to replace it.

III
OTHER THEMES: THE OBLIGATION TO GIVE, THE OBLIGATION TO RECEIVE

To understand completely the institution of 'total services' and of potlatch, one has still to discover the explanation of the two other elements that are complementary to the former. The institution of 'total services' does not merely carry with it the obligation to reciprocate presents received. It also supposes two other obligations just as important: the obligation, on the one hand, to give presents, and on the other, to receive them. The complete theory of these three obligations, of these three themes relating to the same complex, would yield a satisfactory basic explanation for this form of contract among Polynesian clans. For the time being we can only sketch out how the subject might be treated.

It is easy to find many facts concerning the obligation to receive. For a clan, a household, a group of people, a guest, have no option but to ask for hospitality,[34] to receive presents, to enter into trading,[35] to contract alliances, through wives or blood kinship. The Dayaks have even developed a whole system of law and morality based upon the duty one has not to fail to share in the meal at which one is present or that one has seen in preparation.[36]

The obligation to give is no less important; a study of it might enable us to understand how people have become exchangers of goods and services. We can only point out a few facts. To refuse to give,[37] to fail to invite, just as to refuse to accept,[38] is tantamount to declaring war; it is to reject the bond of alliance and commonality.[39] Also, one gives because one is compelled to do so, because the recipient possesses some kind of right of property over anything that belongs to the donor.[40] This ownership is expressed and conceived of as a spiritual bond. Thus in Australia the son-in-law who owes all the spoils of the hunt to his parents-in-law may not eat anything in their presence for fear that their mere breath will poison what he consumes.[41] We have seen earlier the rights of this kind that the taonga nephew on the female side possesses in Samoa, which are exactly comparable to those of the nephew on the female side (vasu) in Fiji.[42]

In all this there is a succession of rights and duties to consume and reciprocate, corresponding to rights and duties to offer and accept. Yet this intricate mingling of symmetrical and contrary rights and duties ceases to appear contradictory if, above all, one grasps that mixture of spiritual ties between things that to some degree appertain to the soul, and individuals, and groups that to some extent treat one another as things.

All these institutions express one fact alone, one social system, one precise state of mind: everything – food, women, children, property, talismans, land, labour services, priestly functions, and ranks – is there for passing on, and for balancing accounts. Everything passes to and fro as if there were a constant exchange of a spiritual matter, including things and men, between clans and individuals, distributed between social ranks, the sexes, and the generations.

IV *good points for your research on Paganism*

NOTE: THE PRESENT MADE TO HUMANS, AND THE PRESENT MADE TO THE GODS

A fourth theme plays a part in this system and moral code relating to presents: it is that of the gift made to men in the sight of the gods and nature. We have not undertaken the general study that would be necessary to bring out its importance. Moreover, the facts we have available do not all relate to those geographical areas to which we have confined ourselves. Finally, the mythological element that we scarcely yet understand is too strong for us to leave it out of account. We shall therefore confine ourselves to a few remarks.

In all societies in Northeast Siberia[43] and among the Eskimos of West Alaska,[44] as with those on the Asian side of the Behring Straits, potlatch[45] produces an effect not only upon men, who vie with one another in generosity, not only upon the things they pass on to one another or consume at it, not only upon the souls of the dead who are present and take part in it, and whose names have been assumed by men, but even upon nature. The exchange of presents between men, the 'namesakes' – the homonyms of the spirits, incite the spirits of the dead, the gods, things, animals, and nature to be 'generous towards them'.[46] The explanation is given that the exchange of gifts produces an abundance of riches. Nelson[47] and Porter[48] have provided us with a good description of these festivals and of their effect on the dead, on wild life, and on the whales and fish that are hunted and caught by the Eskimos. In the kind of language employed by the British trappers they have the expressive titles of 'Asking Festival',[49] or 'Inviting-in Festival'. They normally extend beyond the bounds of the winter villages. This effect upon nature is clearly brought out in one of the recent studies of these Eskimos.[50]

The Asian Eskimos have even invented a kind of contraption, a wheel bedecked with all kinds of provisions borne on a sort of festive mast, itself surmounted by a walrus head. This portion of the mast projects out of the ceremonial tent whose support it forms. Using another wheel, it is manipulated inside the tent and turned in the direction of the sun's movement. The conjunction of all these themes could not be better demonstrated.[51]

It is also evident among the Chukchee[52] and the Koryaka of the far northeast of Siberia. Both carry out the potlatch. But it is the Chukchee of the coast, just like their neighbours, the Yuit, the Asian Eskimos we have just mentioned, who most practise these obligatory and voluntary exchanges of gifts and presents during long drawn-out 'Thanksgiving Ceremonies',[53] thanksgiving rites that occur frequently in winter and that follow one after another in each of the houses. The remains of the banqueting sacrifice are cast into the sea or scattered to the winds; they return to their land of origin, taking with them the wild animals killed during the year, who will return the next year. Jochelson mentions festivals of the same kind among the Koryak, but he has not been present at them, except for the whale festival.[54] Among the latter, the system of sacrifice seems to be very well developed.[55]

Bogoras[56] rightly compares these customs with those of the

Russian *Koliada*: children wearing masks go from house to house demanding eggs and flour that one does not dare refuse to give them. We know that this custom is a European one.[57]

The relationships that exist between these contracts and exchanges among humans and those between men and the gods throw light on a whole aspect of the theory of sacrifice. First, they are perfectly understood, particularly in those societies in which, although contractual and economic rituals are practised between men, these men are the masked incarnations, often Shaman priest-sorcerers, possessed by the spirit whose name they bear. In reality, they merely act as representatives of the spirits,[58] because these exchanges and contracts not only bear people and things along in their wake, but also the sacred beings that, to a greater or lesser extent, are associated with them.[59] This is very clearly the case in the Tlingit potlatch, in one of the two kinds of Haïda potlatch, and in the Eskimo potlatch.

This evolution was a natural one. One of the first groups of beings with which men had to enter into contract, and who, by definition, were there to make a contract with them, were above all the spirits of both the dead and of the gods. Indeed, it is they who are the true owners of the things and possessions of this world.[60] With them it was most necessary to exchange, and with them it was most dangerous not to exchange. Yet, conversely, it was with them it was easiest and safest to exchange. The purpose of destruction by sacrifice is precisely that it is an act of giving that is necessarily reciprocated. All the forms of potlatch in the American Northwest and in Northeast Asia know this theme of destruction.[61] It is not only in order to display power, wealth, and lack of self-interest that slaves are put to death, precious oils burnt, copper objects cast into the sea, and even the houses of princes set on fire. It is also in order to sacrifice to the spirits and the gods, indistinguishable from their living embodiments, who bear their titles and are their initiates and allies.

Yet already another theme appears that no longer needs this

human underpinning, one that may be as ancient as the potlatch itself: it is believed that purchases must be made from the gods, who can set the price of things. Perhaps nowhere is this idea more characteristically expressed than among the Toradja of Celebes Island. Kruyt[62] tells us 'that there the owner must "purchase" from the spirits the right to carry out certain actions on "his" property', which is really theirs. Before cutting "his" wood, before even tilling "his" soil or planting the upright post of "his" house, the gods must be paid. Whereas the idea of purchase even seems very little developed in the civil and commercial usage of the Toradja,[63] on the contrary this idea of purchase from the spirits and the gods is utterly constant.

Malinowski, reporting on forms of exchange that we shall describe shortly, points to acts of the same kind in the Trobriand Islands. An evil spirit, a *tauvau* whose corpse has been found (that of a snake or land crab) may be exorcised by presenting to it one of the *vaygu'a*, a precious object that is both an ornament or talisman and an object of wealth used in the exchanges of the *kula*. This gift has an immediate effect upon the mind of this spirit.[64] Moreover, at the festival of the *mila-mila*,[65] a potlatch to honour the dead, the two kinds of *vaygu'a*, those of the *kula* and those that Malinowski for the first time[66] calls 'permanent' *vaygu'a*, are displayed and offered to the spirits on a platform identical to that of the chief. This makes their spirits benevolent. They carry off to the land of the dead[67] the shades of these precious objects, where they vie with one another in their wealth just as living men do upon returning from a solemn *kula*.[68]

Van Ossenbruggen, who is not only a theorist but also a distinguished observer living on the spot, has noticed another characteristic of these institutions.[69] Gifts to humans and to the gods also serve the purpose of buying peace between them both. In this way evil spirits and, more generally, bad influences, even not personalized, are got rid of. A man's curse allows jealous spirits to enter into you and kill you, and evil influences to act.

Wrongs done to men make a guilty person weak when faced with sinister spirits and things. Van Ossenbruggen particularly interprets in this way the strewing of money along the path of the wedding procession in China, and even the bride-price. This is an interesting suggestion from which a whole series of facts needs to be unravelled.[70]

It is evident that here a start can be made on formulating a theory and history of contract sacrifice. Contract sacrifice supposes institutions of the kind we have described and, conversely, contract sacrifice realizes them to the full, because those gods who give and return gifts are there to give a considerable thing in the place of a small one.

It is perhaps not a result of pure chance that the two solemn formulas of the contract – in Latin, do ut des, in Sanskrit, dadāmi se, dehi me[71] – also have been preserved in religious texts.

NOTE ON ALMS

Later, however, in the evolution of laws and religions, men appear once more, having become again the representatives of the gods and the dead, if they have ever ceased to be. For example, among the Hausa in the Sudan, when the Guinea corn is ripe, fevers may spread. The only way to avoid this fever is to make presents of this grain to the poor.[72] Also among the Hausa (but this time in Tripoli), at the time of the Great Prayer (Baban Salla), the children (these customs are Mediterranean and European) visit houses: 'Should I enter?' The reply is: 'O long-eared hare, for a bone, one gets services.' (A poor person is happy to work for the rich.) These gifts to children and the poor are pleasing to the dead.[73] Among the Hausa these customs may be of Moslem origin,[74] both Negro and European at the same time, and Berber also.

In any case here one can see how a theory of alms can develop. Alms are the fruits of a moral notion of the gift and of fortune[75] on the one hand, and of a notion of sacrifice, on the other. Generosity is an obligation, because Nemesis avenges the poor and the gods for the superabundance of happiness and wealth of certain people who should rid themselves of it. This is the ancient morality of the gift, which has become a principle of justice. The gods and the spirits accept that the share of wealth and happiness that has been offered to them and had been hitherto destroyed in useless sacrifices should serve the poor and children.[76] In recounting this we are recounting the history of the moral ideas of the Semites. The Arab *sadaka* originally meant exclusively justice, as did the Hebrew *zedaqa*:[77] it has come to mean alms. We can even date from the Mischnaic era, from the victory of the 'Poor' in Jerusalem, the time when the doctrine of charity and alms was born, which, with Christianity and Islam, spread around the world. It was at this time that the word *zedaqa* changed in meaning, because in the Bible it did not mean alms.

However, let us return to our main subject: the gift, and the obligation to reciprocate. These documents and comments have not merely local ethnographic interest. A comparison can broaden the scope of these facts, deepening their meaning.

The basic elements of the potlatch[78] can therefore be found in Polynesia, even if the institution in its entirety is not to be found there.[79] In any case 'exchange-through-gift' is the rule there. Yet, it would be merely pure scholasticism to dwell on this theme of the law if it were only Maori, or at the most, Polynesian. Let us shift the emphasis of the subject. We can show, at least as regards the *obligation to reciprocate*, that it has a completely different sphere of application. We shall likewise point out the extension of other obligations and prove that this interpretation is valid for several other groups of societies.

2

THE EXTENSION OF THIS SYSTEM

Liberality, honour, money

THE RULES OF GENEROSITY: THE ANDAMAN ISLANDS*

First, these customs are also to be found among the Pygmies, who, according to Fr Schmidt,[1] are the most primitive of peoples. As early as 1906 Brown observed facts of this kind among the Andaman Islanders (North Island) and described them extremely well with regard to hospitality between local groups and visitors to festivals and fairs that serve as occasions for voluntary and obligatory exchanges (a trade in ochre and sea products against the products of the forest, etc.):

In spite of the considerable volume of these exchanges, since

* See notes for Chapter 2, p. 97, introductory paragraph.

the local group and the family, in other cases, know how to be self-sufficient in tools, etc. . . . these presents do not serve the same purpose as commerce and exchange in more developed societies. The goal is above all a moral one, the object being to foster friendly feelings between the two persons in question, and if the exercise failed to do so, everything had failed.[2]

Nobody is free to refuse the present that is offered. Everyone, men and women, tries to . . . outdo one another in generosity. A kind of rivalry existed to see who could give the greatest number of objects of the greatest value.[3]

Presents put the seal upon marriage and form a link of kinship between the two pairs of parents. They give the two 'sides' the same nature, and this identical nature is made clearly manifest in the prohibition that, from the first betrothal vows to the very end of their days, places a taboo on the two groups of parents, who from then onwards do not see each other or communicate verbally, but continue constantly to exchange presents.[4] In reality this prohibition expresses both the close relations and the fear that reign between this type of reciprocal creditors and debtors. The proof that this is the underlying principle is shown by the fact that the same taboo, indicative simultaneously both of closeness and remoteness in relationships, is also established between young people of both sexes who have undergone at the same time the ceremonies of 'eating turtle and eating pig'[5] and who are likewise bound for life to exchange presents. Facts of this kind are also observed in Australia.[6] Brown again reports on the rituals of meeting after a long separation, the act of embrace, the greeting made in tears, and shows how the exchange of presents is their equivalent,[7] and how feelings and persons are mixed up together.[8]

In short, this represents an intermingling. Souls are mixed with things; things with souls. Lives are mingled together, and this is how, among persons and things so intermingled, each

emerges from their own sphere and mixes together. This is
precisely what contract and exchange are.

II
PRINCIPLES, REASONS, AND THE INTENSITY OF EXCHANGE OF GIFTS (MELANESIA)

More so than the peoples of Polynesia, those of Melanesia have
preserved or developed the potlatch,[9] although this is not a
matter that concerns us here. In any case, the Melanesians, better
than the Polynesians, have on the one hand preserved, and on
the other, developed the whole system of gifts and this form of
exchange. Since, moreover, with the former the notion of
money[10] emerges much more clearly than in Polynesia, the
system becomes in part complicated, but is also more clear-cut.

New Caledonia

Here we again find not only those ideas we seek to highlight, but
even their expression in the characteristic documents that
Leenhardt has collected about the New Caledonians. He began
by describing the pilou-pilou and the system of festivals, gifts and
services of all kinds[11] that we should not hesitate to term 'potlatch'. The legal terms used in the solemn speeches made by the
herald are entirely typical. Thus, at the ceremonial presentation
of festival yams,[12] the herald says: 'If there is some ancient pilou
before which we have not been, there, among the Wi, etc. . . . ,
this yam hastens to it as once such a yam came from them to
us.'[13] It is the thing itself that comes back. Later in the same
speech, it is the spirit of their ancestors who causes 'to descend . . .
upon these portions of food the effects of their action and
strength.' 'The result of the action you have accomplished
appears today. Every generation has appeared in its mouth.'
Another way of representing the legal tie, one no less expressive,

is: 'Our festivals are the movement of the hook that serves to bind together the various sections of the straw roofing so as to make one single roof, one single word.'[14] It is the same things that return, the same thread that passes through.[15] Other authors also point out these facts.[16]

Trobriand Islands

At the other end of the Melanesian world a very well-developed system is equivalent to that of the New Caledonians. The inhabitants of the Trobriand Islands are among the most civilized of these races. Today they are wealthy pearl fishermen, and, before the arrival of the Europeans, they were rich pottery manufacturers and makers of shell money, stone axes, and precious goods. They have always been good traders and bold navigators. Malinowski gives them a name that fits them exactly when he compares them to Jason's fellow voyagers: 'Argonauts of the Western Pacific'. In one of the best volumes of descriptive sociology, focusing, so to speak, on the subject that concerns us, he has described the complete system of inter- and intratribal trade that goes under the name of *kula*.[17] We still await from him the description of all the institutions that are governed by the same principles of law and economics: marriage, the festival of the dead, initiation, etc. Consequently, the description that we shall give is still only provisional. But the facts are of capital importance, and are plain.[18]

The *kula* is a sort of grand potlatch. The vehicle for busy intertribal trade, it extends over the whole of the Trobriand Islands, a part of the Entrecasteaux Islands, and the Amphlett Islands. In all these territories it indirectly involves all the tribes and, directly, a few of the large tribes – the Dobu in the Amphletts, the Kiriwina, the Sinaketa, and Kitav in the Trobriands, and the Vakuta on Woodlark Island. Malinowski gives no translation of *kula*, which doubtless means 'circle'. Indeed it is as if all these tribes, these expeditions across the sea, these precious things and objects for use, these types of food and festivals, these services rendered of all kinds, ritual and sexual, these men and women, – were caught up in a circle,[19] following around this circle a regular movement in time and space.

Kula trade is of a noble kind.[20] It seems to be reserved for the chiefs. The latter are at one and the same time the leaders of fleets of ships and boats. They are the traders, and also the recipients of gifts from their vassals, who are in fact also their children and brothers-in-law, their subjects, and at the same time the chiefs of various vassal villages. Trade is carried on in a noble fashion, apparently in a disinterested and modest way.[21] It is distinguished carefully from the mere economic exchange of useful goods, which is called *gimwali*.[22] In fact, the latter is carried on, as well as the *kula*, in the large primitive fairs that constitute the gatherings of the intertribal *kula*, or in the small markets of the intratribal *kula*. It is marked by very hard bargaining between the two parties, a practice unworthy of the *kula*. Of an individual who does not proceed in the *kula* with the necessary greatness of soul, it is said that he is 'conducting it like a *gimwali*.' In appearance, at the very least, the *kula* – as in the potlatch of the American Northwest – consists in giving by some, and receiving by others.[23] The recipients of one day become the givers on the next. In the most complete form, the most solemn, lofty, and competitive form of the *kula*,[24] that of the great sea expeditions, the *Uvalaku*, it is even the rule to leave without having anything to exchange, without even having anything to give, although it might be exchanged for food, which one refuses even to ask for. One pretends only to receive. It is when the visiting tribe plays host the following year to the fleet of the tribe that has been visited that the presents will be reciprocated with interest.

However, in *kula* not given on such a grand scale, advantage is taken of the sea journey to exchange cargoes. The nobles themselves carry on trade. About this there is much native theory.

Numerous objects are solicited,[25] asked for, and exchanged, and every kind of relationship is established outside the *kula*, which, however, always remains the purpose, and the decisive moment in these relationships.

The act of giving itself assumes very solemn forms: the thing received is disclaimed and mistrusted; it is only taken up for a moment, after it has been cast at one's feet. The giver affects an exaggerated modesty:[26] having solemnly brought on his present, to the sound of a seashell, he excuses himself for giving only the last of what remains to him, and throws down the object to be given at the feet of his rival and partner.[27] However, the seashell and the herald proclaim to everybody the solemn nature of this act of transfer. The aim of all this is to display generosity, freedom, and autonomous action, as well as greatness.[28] Yet, all in all, it is mechanisms of obligation, and even of obligation through things, that are called into play.

The essential objects in these exchange-gifts are the *vaygu'a*, a kind of money.[29] It is of two kinds: the *mwali*, which are beautiful bracelets, carved, polished, and placed in a shell, and worn on great occasions by their owners or relatives; and the *soulava*, necklaces fashioned by the skilful craftsmen of Sinaketa in a pretty mother-of-pearl made from red spondylus. They are solemnly worn by the women,[30] and, in cases of great anguish, exceptionally by the men.[31] Normally, however, both kinds are hoarded and treasured. They are kept for the sheer pleasure of possessing them. The making of the bracelets, fishing for and making the necklaces into jewellery, the trade in these two objects of exchange and prestige, together with other forms of trade that are more profane and vulgar, constitute the source of the Trobriand people's fortune.

According to Malinowski, these *vaygu'a* follow a kind of circular movement: the *mwali*, the bracelets, are passed on regularly from west to east, whereas the *soulava* always travel from east to west.[32] These two movements in opposite directions occur in all the islands – Trobriand, Entrecasteaux, Amphlett, the remote islands – Woodlark, Marshall Bennett, Tubetube – and finally the extreme southeast coast of New Guinea, from which come the unworked bracelets. There this trade meets the great expeditions of the same kind that come from New Guinea (South Massim),[33] which Seligmann has described.

In principle the circulation of these signs of wealth is continuous and unerring. They must not be kept too long a time, nor must one be slow or difficult in passing them on.[34] One should not present them to anyone other than certain partners, nor save in a certain direction – the 'bracelet' or the 'necklace' direction.[35] One can and should keep them from one kula to the next, and the whole community is proud of the vaygu'a that one of its chiefs has obtained. There are even occasions, such as in the preparation of funeral ceremonies, of great s'oi, when it is permitted always to receive and to give nothing in return.[36] Yet this is in order to give back everything and to spend everything, when the festival has begun. Thus it is indeed ownership that one obtains with the gift that one receives. But it is ownership of a certain kind. One could say that it partakes of all kinds of legal principles that we, more modern, have carefully isolated from one another. It is ownership and possession, a pledge and something hired out, a thing sold and bought, and at the same time deposited, mandated, and bequeathed in order to be passed on to another. For it is only given you on condition that you make use of it for another or pass it on to a third person, the 'distant partner', the murimuri.[37] Such is the nature of this economic, legal, and moral entity, which is truly typical, as Malinowski was able to discover, find again, observe, and describe.

This institution has also its mythical, religious, and magical aspect. The vaygu'a are not unimportant things, mere pieces of money. Each one, at least the dearest and the most sought after – and other objects enjoy the same prestige[38] – each one has its name,[39] a personality, a history, and even a tale attached to it. So much is this so that certain individuals even take their own name from them. It is not possible to say whether they are really the object of a cult, for the Trobriand people are, after their fashion, positivists. Yet one cannot fail to acknowledge the eminent and sacred nature of the objects. To possess one is 'exhilarating, strengthening, and calming in itself.'[40] Their owners fondle and look at them for hours. Mere contact with them passes on their virtues.[41] *Vaygu'a* are placed on the forehead, on the chest of a dying person, they are rubbed on his stomach, and dangled before his nose. They are his supreme comfort.

Yet there is even more to it than this. The contract itself partakes of this nature of the *vaygu'a*. Not only the bracelets and the necklaces, but even all the goods, ornaments, and weapons, everything that belongs to the partner, is so imbued with it, at least emotionally if not in his inmost soul, that they participate in the contract.[42] A very beautiful phrase, 'the enchantment of the seashell'[43] serves, after the possessions have been evoked, to cast a spell over and move towards the 'partner-candidate' the things that he must ask for and receive.[44]

> [A state of excitement[45] takes hold of my partner][46]
> A state of excitement takes hold of his dog,
> A state of excitement takes hold of his belt . . .

And so on: 'of his *gwara* [the taboo on coconuts and betel nuts];[47] of his necklace *bagido'u*; of his necklace *bagiriku*; of his necklace *bagidudu*,[48] etc., etc.'

Another more mythical,[49] even more curious phrase, but of a more common type, expresses the same idea. The partner of the *kula* has an animal to assist him, a crocodile that he calls upon that has to bring him the necklaces [in Kitava, the *mwali*]:

> The crocodile falls upon him, carries off your man, and shoves him under the *gebobo* (the cargo hold on a boat).

> Crocodile, bring me the necklace, bring me the *bagido'u*, the *bagiriku*, etc.

A formula that comes earlier in the same ritual invokes a bird of prey.[50]

The last formula of enchantment used by those associated with or contracting in the ritual [at Dobu or at Kitava, by the people of Kiriwina] contains a couplet[51] of which two interpretations are given. Moreover, the ritual is very long and is repeated at length. Its purpose is to enumerate all that the *kula* proscribes, all the things relating to hatred and war that must be exorcised in order to be able to trade between friends.

> Your fury, the dog turns up its nose at it;
> Your war paint, the dog turns up its nose at it, etc.,

Other versions go as follows:[52]

> Your fury, the dog is docile at, etc.

or:

> Your fury takes off like the tide. The dog plays;
> Your anger takes off like the tide. The dog plays; etc.

This must be understood as: 'Your fury becomes like the dog who is playing.' The essential element is the metaphor of the dog that gets up to lick the hand of its master. So the man, if not the woman of Dobu, should also act. A second interpretation, sophisticated and not untinged with scholasticism, so Malinowski declares, but clearly a very local one, gives a different gloss that ties in better with what we know:

> The dogs are playfully nuzzling one another. When you

mention this word 'dog', the precious things also come [to play], as has long been ordained. We have given bracelets, necklaces will come. Both will meet each other (like the dogs who come sniffing at one another).

The expression, in the form of a parable, is a pretty one. The entire set of collective sentiments is expressed at a stroke: the potential hatred between associates, the isolation of the *vaygu'a*, ceasing as if by magic; men and precious things coming together like dogs that play and run up at the sound of one's voice.

Another symbolic expression is that of the marriage of the *mwali*, the bracelets, the feminine symbols, with the *soulava*, the necklaces, the masculine symbols, which stretch out towards each another, as does male towards female.[53]

These various metaphors signify exactly the same thing as is characterized in different terms by the mythical jurisprudence of the Maori. Sociologically, it is once again the mixture of things, values, contracts, and men that is so expressed.[54]

Unfortunately, our knowledge of the legal rule that governs these transactions is defective. It is either an unconscious rule, imperfectly formulated by the Kiriwina people, Malinowski's informants; or, if it is clear for the Trobriand people, it should be the subject of a fresh enquiry. We only possess details. The first gift of a *vaygu'a* bears the name of *vaga*, 'opening gift'.[55] It is the starting point, one that irrevocably commits the recipient to make a reciprocating gift, the *yotile*,[56] which Malinowski felicitously translates as the 'clinching gift': the gift that seals the transaction. Another name for this latter gift is *kudu*, the tooth that bites, that really cuts, bites through, and liberates.[57] It is obligatory; it is expected, and it must be equivalent to the first gift. Occasionally it may be seized by force or by surprise.[58] For a *yotile* that is an insufficient return gift, revenge may[59] be taken[60] by magic, or at the very least by insult and a display of resentment. If one is not able to reciprocate, at the very least one may offer a *basi*, which merely 'pierces' the skin, does not bite, and does not conclude the affair. It is a kind of advance present, whose purpose is to delay. It appeases the former donor, now the creditor; but does not free the debtor,[61] the future donor. These are all curious details, and everything about these expressions is striking. Yet we do not know the sanction behind it. Is it purely moral[62] and magical? The individual who is 'obdurate at the *kula*', is he only scorned, and if needs be, cast under a spell? Does the partner who does not keep faith lose anything else: his noble rank, or at least his place among the chiefs? This we still need to know.

Yet from another viewpoint the system is typical. Except for ancient Germanic law that we shall be discussing later, in the present state of our observations and historical, juridical, and economic knowledge, it would be difficult to come across a custom of gift-through-exchange more clear-cut, complete, and consciously performed, and, moreover, better understood by the observer recording it than the one Malinowski found among the Trobriand people.[63]

The *kula*, its essential form, is itself only one element, the most solemn one, in a vast system of services rendered and reciprocated, which indeed seems to embrace the whole of Trobriand economic and civil life. The *kula* seems to be merely the culminating point of that life, particularly the *kula* between nations and tribes. It is certainly one of the purposes of existence and for undertaking long voyages. Yet in the end, only the chiefs, and even solely those drawn from the coastal tribes – and then only a few – do in fact take part in it. The *kula* merely gives concrete expression to many other institutions, bringing them together.

First, the exchange of the *vaygu'a* themselves during the *kula* forms the framework for a whole series of other exchanges, extremely diverse in scope, ranging from bargaining to remuneration, from solicitation to pure politeness, from out-and-out hospitality to reticence and reserve. In the first place, except for the *uvalaku*, large-scale solemn expeditions that are purely ceremonial and competitive,[64] all the *kula* provide the occasion for *gimwali*, which are commonplace exchanges, not necessarily occurring between partners.[65] A free market exists between individuals of allied tribes, side by side with closer associations. In the second place, between partners in a *kula* there pass, as if in a perpetual chain, additional gifts, presented and reciprocated, as well as obligatory transactions. The *kula* even takes these for granted. The association that is constituted, and which is the principle of the *kula*,[66] begins with a first gift, the *vaga*, that is solicited with all one's might by means of 'inducements'. For this first gift the future partner, still a free agent, can be wooed, and he is rewarded, so to speak, by a preliminary series of gifts.[67] Whilst one is certain that the reciprocating *vaygu'a*, the *yotile*, the 'clinching gift', will be returned, one is not sure that the *vaga* will be given, or even that the 'inducements' will be accepted. This way of soliciting and accepting a gift is the rule; each of the presents made in this way bears a special name. They are placed on display before being offered. In that case they are termed *pari*.[68] Others bear names that indicate the noble and magical nature of the object that is offered.[69] But if one accepts one of these presents it shows that one is disposed to enter into the game, if not to remain in it for long. Certain names given to these presents express the legal situation that acceptance of them entails:[70] this time the matter is regarded as settled. The present is normally something fairly valuable: for example, a large, polished stone axe, or a whalebone spoon. To accept it is to bind oneself definitively to making a gift of the *vaga*, the first present that is sought after. But one is still only half-committed as a partner. Only the solemn observance of the tradition commits one completely. The importance and nature of these gifts springs from the extraordinary competition that occurs between the potential partners in the expedition that arrives. They seek out the best possible partner in the opposing tribe. The affair is a serious matter, for the association one attempts to create establishes a kind of clan link between the partners.[71] Thus to choose, one must attract and dazzle the other person.[72] Whilst rank is taken into account,[73] one must attain one's goal before the others, or in a better way than they do, so bringing about more plentiful exchanges of the most valuable things, which are naturally the property of the richest people. Competition, rivalry, ostentatiousness, the seeking after the grandiose, and the stimulation of interest – these are the various motives that underlie all these actions.[74]

These are the arrival gifts. Other gifts, equivalent in value to them, are made in return. These are the departure gifts (called *talo'i* at Sinaketa),[75] made upon taking leave; they are always superior to the arrival gifts. Already the cycle of total services and total counter-services with interest is being completed side by side with the *kula*.

There have naturally been – throughout the time these transactions last – services of hospitality and food, and at Sinaketa, relating to women.[76] Finally, throughout this period, other additional gifts, always reciprocated regularly, are offered. It would even appear to us that the exchange of these *korotumna* represents a primitive form of the *kula* – when it consisted also of exchanging stone axes[77] and the rounded tusks of the wild pig.[78]

Moreover, in our view the whole intertribal *kula* is merely the extreme case, the most solemn and most dramatic one, of a more general system. This takes the tribe itself, in its entirety, out of the narrow sphere of its physical boundaries, and even of its interests and rights. Yet within the tribe the clans and villages are normally joined by links of the same kind. In that case it is only local and domestic groups, together with their chiefs, who leave their homes, pay visits, trade, and intermarry. This may no longer be termed a *kula*. However, Malinowski, contrasting it with the 'coastal *kula*', rightly talks of the 'internal *kula*' and of 'kula communities', which provide the chief with the objects he will exchange. However, it is not overstating it to speak in such cases of a potlatch proper. For example, the visits of the Kiriwana people to Kitawa for the s'oi,[79] the funeral festivals, include many things other than the exchange of the vaygu'a. One can see them as a kind of simulated attack (youlawada),[80] and a distribution of food, with a display of pigs and yams.

Furthermore, the vaygu'a and all such objects are not always acquired, made, and exchanged by the chiefs themselves.[81] Indeed, it may be said that they are neither made[82] nor exchanged by the chiefs for their own advantage. Most come to the chiefs in the form of gifts from relatives of a lower rank, in particular from brothers-in-law, who are at the same time vassals,[83] or from sons who hold land separately as vassals. In return, most of the vaygu'a, upon the return of the expedition, are solemnly passed on to the chiefs of the villages and clans, and even to the common people of associated clans – in brief, to whoever has taken part, directly or indirectly, and often very indirectly, in the expedition.[84] In this way the latter are compensated.

Finally, side by side with, or if one wishes, above, below, and all around, and, in our opinion, at the bottom of this system of internal kula, the system of gift-through-exchange permeates all the economic, tribal, and moral life of the Trobriand people. It is 'impregnated' with it, as Malinowski very neatly expressed it. It is a constant 'give and take'.[85] The process is marked by a continuous flow in all directions of presents given, accepted, and reciprocated, obligatorily and out of self-interest, by reason of greatness and for services rendered, through challenges and pledges. Here we cannot set out all the facts. Moreover, their publication by Malinowski himself is not yet complete. First, however, here are two main facts.

An entirely analogous relationship to that of the kula is that of the wasi.[86] It establishes regular acts of exchange, which are obligatory between, on the one hand, agricultural tribes, and on the other hand, coastal tribes. The agricultural partner comes to lay his products in front of the house of his fisherman associate. On another occasion, the latter, after a big fishing expedition, will go to the agricultural village to repay these with interest from the fruits of his catch.[87] It is the same system of division of labour as we have noted in New Zealand.

Another important kind of exchange takes on the form of exhibitions.[88] Such are the *sagali*, distributions of food[89] on a grand scale, that are made on several occasions: at harvest time, at the building of the chief's hut or new boats, or at funeral festivals.[90] These distributions are made to groups that have performed some service for the chief or his clan:[91] cultivation of the land, the transporting of the large tree trunks from which boats or beams are carved, and for services rendered at funerals by the members of the dead person's clan, etc. These distributions are absolutely equivalent to the Tlingit potlatch. Even the themes of combat and rivalry appear. In it, clans and phratries, and families allied to one another, confront one another. Generally distributions seem to be due to group action, in so far as the personality of the chief does not make itself felt.

Yet in addition to these group rights and this collective economy, already less resembling the *kula*, all individual relationships of exchange seem to us to be of this type. Perhaps only a few consist of mere barter. However, as barter is hardly carried on except between relatives, allies, or partners in the *kula* and the *wasi*, it does not seem that exchange is really free. Generally, even what is received and has come into one's possession in this way – in whatever manner – is not kept for oneself, unless one cannot do without it. Normally it is passed on to someone else, to a brother-in-law, for example.[92] It can happen that the identical things one has acquired and then given away come back to one in the course of the same day.

All the rewards for 'total services' of any kind, things and services, fall within this category. In random order, the following are the most important.

The pokala[93] and kaributu,[94] 'solicitory gifts', which we have noted in the kula, are species of a much larger category that corresponds fairly closely to what we term 'remuneration'. They are offered to the gods and the spirits. Another generic name for the 'remuneration' is vakapula[95] and mapula.[96] They are marks of gratitude and hospitable welcome and must be reciprocated. In this connection, in our opinion Malinowski[97] has made a very great discovery that sheds light upon all the economic and juridical relationships between the sexes within marriage: the services of all kinds rendered to the wife by her husband are considered as a remuneration-cum-gift for the service rendered by the wife when she lends what the Koran still calls 'the field'.

The somewhat childish legal language of the Trobriand Islanders has given rise to a proliferation of distinctive names for all kinds of total counter-services, according to the name of the service that is being compensated,[98] the thing given,[99] the occasion,[100] etc. Certain names take all these considerations into account; for example, the gift made to a magic man, or for the acquisition of a title, is called laga.[101] One cannot credit the extent to which all such vocabulary is complicated by a curious incapacity to divide and define, and by the strange refinements that are given to names.

Other Melanesian societies

One need not multiply the comparisons with other areas of Melanesia. However a few details gleaned here and there will strengthen conviction and prove that the Trobriand people and the New Caledonians have not developed in an abnormal way a principle that might not be found among kindred peoples.

At the southern limit of Melanesia, in Fiji, where we have identified the existence of potlatch, other remarkable institutions thrive that belong to the gift system. There is a season, termed *kere-kere*, during which nobody must be refused anything.[102] Gifts are exchanged between the two families on the occasion of a marriage, etc.[103] Moreover the money of Fiji, of sperm whale's teeth, is exactly of the same kind as that of the Trobriands. It bears the name of *tambua*;[104] it is decorated with stones ('mothers of the teeth') and ornaments that are kinds of mascots, talismans, and 'good-luck' objects of the tribe. The feelings cherished by the Fijians for their *tambua* are exactly the same as those we have just described: 'they are treated like dolls. They are taken out of the basket and admired, their beauty is spoken of; their "mother" is oiled and polished.'[105] To present them constitutes a request; to accept them is to commit oneself.[106]

The Melanesians of New Guinea and certain Papuans influenced by them call their money *tau-tau*.[107] It is of the same kind and the object of the same beliefs as that of the Trobriand Islanders.[108] But this name must also be compared with that of *tahu-'ahu*,[109] which means the 'borrowing of pigs' (Motu and Koita), and is a name[110] familiar to us. It is the very Polynesian term, the root of the word *taonga*, which in Samoa and New Zealand means the jewels and possessions incorporated into the family. The words themselves are Polynesian, as are the things.[111]

It is known that the Melanesians and Papuans of New Guinea practise the potlatch.[112]

The fine documents that Thurnwald has passed on to us concerning the tribes of Buin[113] and the Banaro[114] have already provided us with numerous points of comparison. The religious nature of the things exchanged is apparent, particularly that of money, in the way that it rewards songs, women, love, and services. As in the Trobriand Islands, it is a kind of pledge. Finally, Thurnwald has analysed in a sound case-study[115] one of the facts that best illustrates both what this system of reciprocal gifts consists of, and what is incorrectly called marriage by purchase. In reality this seems to include services to all involved, including the family-in-law. The wife whose parents have not sent sufficient return presents is sent back to them.

To sum up, the whole area of the islands, and probably part of the world of Southern Asia that is related to it, possess the same legal and economic system.

The conception one should have regarding these Melanesian tribes, even richer and more committed to trade than the Polynesians, is therefore very different from normal. These peoples possess an extra domestic economy and a very developed system of exchange that throbs with life more intensely and more precipitantly perhaps than the one that our peasants or the fishing villages along our coasts were familiar with maybe not even a hundred years ago. They have an extensive economic life, going beyond the confines of the islands and their dialects, which represents a considerable trade. Through gifts made and reciprocated they have robustly replaced a system of buying and selling.

The stumbling block that these laws and, as we shall also see, Germanic law, came up against, was their inability to isolate and divide up their economic and juridical concepts. But they had no need to do so. In these societies neither the clan nor the family is able to isolate itself or dissociate its actions. Nor are individuals themselves, however influential and aware, capable of understanding that they need to oppose one another and learn how to dissociate their actions from one another. The chief is merged with his clan, and the clan with him. Individuals feel themselves acting in only one way. Holmes perceptively remarks that the two dialects, Papuan and Melanesian, of the tribes that he encountered at the mouth of the Finke river (Toaripi and Namau) have 'one single term to designate buying and selling, lending and borrowing'. Operations that are 'opposites are expressed by the same word'.[116] 'Strictly speaking they did not know how to borrow and lend in the sense that we employ these terms, but there was always something given in the shape of an honorarium for borrowing, and which was returned when the loan was repaid.'[117] These men have no conception of either selling or borrowing, but nevertheless carry out juridical and economic operatons that fulfil the same functions.

Likewise the notion of barter is no more natural for the Melanesians than for the Polynesians.

One of the best ethnographers, Kruyt, whilst he uses the word 'selling', describes exactly[118] this state of mind as it exists among the inhabitants of the central Celebese islands. Yet the Toradja have been for a long while in contact with the Malaysians, who are great traders.

Thus one section of humanity, comparatively rich, hard-working, and creating considerable surpluses, has known how to, and still does know how to, exchange things of great value, under different forms and for reasons different from those with which we are familiar.

III
THE AMERICAN NORTHWEST

Honour and credit

From these observations about a few Melanesian and Polynesian peoples there already emerges a very clearly defined picture of this system of the gift. Material and moral life, and exchange, function within it in a form that is both disinterested and obligatory. Moreover, this obligation is expressed in a mythical and imaginary way or, one might say, symbolic and collective. It assumes an aspect that centres on the interest attached to the things exchanged. These are never completely detached from those carrying out the exchange. The mutual ties and alliance that they establish are comparatively indissoluble. In reality this symbol of social life – the permanence of influence over the things exchanged – serves merely to reflect somewhat directly the manner in which the subgroups in these segmented societies, archaic in type and constantly enmeshed with one another, feel that they are everything to one another.

The Indian societies of the American Northwest display the same institutions, although with them they are even more radical and more marked. First, one can say that barter is unknown. Even after long contact with Europeans,[119] apparently none of the considerable transfers of wealth[120] constantly taking place among them is carried out save in the solemn form of the potlatch.[121] We shall describe this institution as it relates to our study.

Before doing so, a brief description of these societies is indispensable. The tribes, peoples, or rather groups of tribes[122] we shall discuss all reside on the Northwest coast of America, in Alaska: Tlingit and Haïda; and in British Columbia, mainly the Haïda, Tsimshian, and Kwakiutl.[123] They also live from the sea, or from the rivers, from fishing rather than hunting. But, unlike the Melanesians and Polynesians, they have no agriculture. However, they are very rich, and even now their fishing grounds, hunting grounds, and fur-trapping provide them with considerable surpluses, particularly when reckoned in European terms. They have the most solidly built houses of all the American tribes, and a very highly developed cedarwood industry. Their boats are good, and although they hardly venture out on the open sea, they know how to navigate between the islands and the coasts. Their material arts are of a very high order. In particular, even before the arrival of iron in the eighteenth century, they knew how to extract, melt down, mould, and beat out the copper that is to be found in a raw state in the Tsimshian and Tlingit lands. Certain kinds of this copper, real armorial shields, served as a kind of money for them. Another kind of money was certainly the beautiful, so-called Chilkat, blankets of wonderfully different colours that still serve as adornment,[124] some of them of considerable value. These peoples have excellent sculptors and professional designers. Their pipes, tomahawks, sticks, spoons carved out of horn, etc., embellish our ethnographic collections. The whole of this civilization is remarkably uniform, within very broad limits. Clearly, these societies mingled with one another from very ancient times, although they belong, at least in language, to no less than three different families of peoples.[125] Their life in winter, even for the southernmost tribes, is very different from that in summer. The tribes have a dual structure: from the end of spring they disperse to go hunting, to gather roots and the juicy mountain berries, and to fish for salmon in the rivers; at the onset of winter they concentrate once more in what are called 'towns'. It is then, during the period when they are gathered together in this way, that they live in a state of perpetual excitement. Social life becomes extremely intense, even more so than in the assemblies of tribes that can take place in the summer. There are constant visits from whole tribes to tribes, from clans to clans, and from families to families. There are repeated festivals, continuous and long drawn-out. At a wedding, or at various kinds of ritual or promotions, everything stored up with great industry during the summer and autumn on one of the richest coasts in the world is lavishly expended. This even occurs in domestic life. The people of one's clan are invited when a seal has been killed or when a case of berries or roots that have been preserved is opened up. Everyone is invited when a whale is washed up. From the moral viewpoint also, the civilization is remarkably uniform, although ranging from the regime of the phratry (Tlingit and Haïda) of maternal descent, to the clan of modified masculine descent of the Kwakiutl. The general features of social organization, and particularly totemism, are roughly the same in all the tribes. In the Banks Islands the tribes have 'brotherhoods', as in Melanesia, which are wrongly called secret societies and are often international, but in which male society, and certainly among the Kwakiutl, female society, cuts across the clan organization. Part of the gifts and total counter-services of which we shall speak are intended, as in

Melanesia,[126] to finance the ranks and successive 'ascensions' (promotions)[127] in these brotherhoods. The rituals, both of these brotherhoods and clans, follow one another at the marriage of chiefs, at 'copper sales', at initiations, at Shamon ceremonies, and at funeral ceremonies – the latter being more developed in Haïda and Tlingit lands. All are performed during a series of potlatches that are prolonged indefinitely. There are potlatches everywere, in response to other potlatches. As in Melanesia it is a constant 'give and take'.

The potlatch itself, so typical a phenomenon, and at the same time so characteristic of these tribes, is none other than the system of gifts exchanged.[128] It differs from that in Melanesia only in the violence, exaggeration, and antagonisms that it arouses, and by a certain lack of juridical concepts, and a simpler and cruder structure. This is particularly true of the two northern peoples, the Tlingit and Haïda.[129] The collective nature of the contract[130] is more apparent than in Melanesia and Polynesia. All in all, these societies, in spite of appearances, are closer to what we term 'simple total services'. Consequently, the juridical and economic concepts possess less clarity and less conscious precision. However, in practice, the principles are positive and sufficiently clear-cut.

Nevertheless, two notions are much more in evidence than in the Melanesian potlatch or the more developed, or more fragmented, institutions existing in Polynesia. These are the notion of credit, of the time limit placed on it, and also the notion of honour.[131]

Gifts circulate, as we have seen in Melanesia and Polynesia, with the certainty that they will be reciprocated. Their 'surety' lies in the quality of the thing given, which is itself that surety. But in every possible form of society it is in the nature of a gift to impose an obligatory time limit. By their very definition, a meal shared in common, a distribution of *kava*, or a talisman that one takes away, cannot be reciprocated immediately. Time is needed in order to perform any counter-service. The notion of a time limit is thus logically involved when there is question of returning visits, contracting marriages and alliances, establishing peace, attending games or regulated combats, celebrating alternative festivals, rendering ritual services of honour, or 'displaying reciprocal respect'[132] – all the things that are exchanged at the same time as other things that become increasingly numerous and valuable, as these societies become richer.

Current economic and juridical history is largely mistaken in this matter. Imbued with modern ideas, it forms *a priori* ideas of development[133] and follows a so-called necessary logic. All in all, it rests on ancient traditions. There is nothing more dangerous than this 'unconscious sociology', as Simiand has termed it. For example Cuq still states:

> In primitive societies only the barter regime is conceived of; in those more advanced, sales for cash are the practice. Sale on credit is characteristic of a higher phase in civilization. It first appears in an oblique form as a combination of cash sale and loans.[134]

(Cuq 1910)

In fact, the point of departure lies elsewhere. It is provided in a category of rights that excludes the jurists and economists, who are not interested in it. This is the gift, a complex phenomenon, particularly in its most ancient form, that of 'total services', with which we do not deal in this monograph. Now, the gift necessarily entails the notion of credit. The evolution in economic law has not been from barter to sale, and from cash sale to credit sale. On the one hand, barter has arisen through a system of presents given and reciprocated according to a time limit. This was through a process of simplification, by reductions in periods of time formerly arbitrary. On the other hand, buying and selling arose in the same way, with the latter according to a fixed time limit, or by cash, as well as by lending. For we have no evidence that any of the legal systems that have evolved beyond the phase we are describing (in particular, Babylonian law) remained ignorant of the credit process that is known in every archaic society that still survives today. This is another simple, realistic way of resolving the problem of the two 'moments in time' that are brought together in the contract, which Davy has already studied.[135]

No less important in these transactions of the Indians is the role played by the notion of honour. Nowhere is the individual prestige of a chief and that of his clan so closely linked to what is spent and to the meticulous repayment with interest of gifts that have been accepted, so as to transform into persons having an obligation those that have placed you yourself under a similar obligation. Consumption and destruction of goods really go beyond all bounds. In certain kinds of potlatch one must expend all that one has, keeping nothing back.[136] It is a competition to see who is the richest and also the most madly extravagant. Everything is based upon the principles of antagonism and rivalry. The political status of individuals in the brotherhoods and clans, and ranks of all kinds, are gained in a 'war of property',[137] just as they are in real war, or through chance, inheritance, alliance, and marriage. Yet everything is conceived of as if it were a 'struggle of wealth'.[138] Marriages for one's children and places in the brotherhoods are only won during the potlatch, where exchange and reciprocity rule. They are lost in the potlatch as they are lost in war, by gambling[139] or in running and wrestling.[140] In a certain number of cases, it is not even a question of giving and returning gifts, but of destroying,[141] so as not to give the slightest hint of desiring your gift to be reciprocated. Whole boxes of olachen (candlefish) oil or whale oil[142] are burnt, as are houses and thousands of blankets. The most valuable copper objects are broken and thrown into the water, in order to put down and to 'flatten' one's rival.[143] In this way one not only promotes oneself, but also one's family, up the social scale. It is therefore a system of law and economics in which considerable wealth is constantly being expended and transferred. If one so wishes, one may term these transfers acts of exchange or even of trade and sale.[144] Yet such trade is noble, replete with etiquette and generosity. At least, when it is carried on in another spirit, with a view to immediate gain, it becomes the object of very marked scorn.[145]

As may be seen, the notion of honour, which expresses itself violently in Polynesia and is always present in Melanesia, is, in this case, really destructive. Again on this point, conventional wisdom misjudges the importance of the motives that have inspired men, and all we owe to the societies that have preceded us. Even such a perceptive scholar as Huvelin felt himself obliged to deduce the notion of honour, reputedly ineffective, from the notion of the efficacy of magic.[146] He sees in honour and prestige only a substitute for magic. The reality is more complex. The notion of honour is no more foreign to these civilizations than is the notion of magic.[147] The Polynesian word *mana* itself symbolizes not only the magical force in every creature, but also his honour, and one of the best translations of the word is 'authority', 'wealth'.[148] The Tlingit and Haïda potlatch consists of considering the mutual services rendered as acts of honour.[149] Even among really primitive tribes, such as those in Australia, the point of honour is as delicate as that in our own societies, and is satisfied through the offering of services and food, acts of precedence and rites, as well as through gifts.[150] Men had learnt how to pledge their honour and their name long before they knew how to sign the latter.

The North American potlatch has been well enough studied as regards everything concerning the form of the contract itself. However, we need to fit the study of it made by Davy and Leonhard Adam[151] into the wider context in which it should be placed for the subject with which we are dealing. For the potlatch is much more than a juridical phenomenon: it is one that we propose to call 'total'. It is religious, mythological, and Shamanist, since the chiefs who are involved represent and incarnate their ancestors and the gods, whose name they bear, whose dances they dance and whose spirits possess them.[152] The potlatch is also an economic phenomenon, and we must gauge the value, the importance, the reasons for, and the effect of these transactions, enormous even today, when they are calculated in European values.[153] The potlatch is also a phenomenon of social structure: the gathering together of tribes, clans, and families, even of peoples, brings about a remarkable state of nerviness and excitement. One fraternizes, yet one remains a stranger; one communicates and opposes others in a gigantic act of trade and a constant tournament.[154] We pass over the aesthetic phenomena, which are extremely numerous. Finally, even from the juridical viewpoint, to what we have already gleaned regarding the form of these contracts and what might be termed their human purpose, as well as the juridical status of the contracting parties (clans, families, ranks, and betrothals,) we must add this: the material purposes of the contracts, the things exchanged in them, also possess a special intrinsic power, which causes them to be given and above all to be reciprocated.

It would have been useful – if we had enough space – to distinguish, for the purposes of our exposition, between four forms of the potlatch in the American Northwest: (1) a potlatch in which the phratries and the families of chiefs are exclusively, or almost exclusively involved (Tlingit); (2) a potlatch in which the phratries, clans, chiefs, and families play roughly an equal part; (3) a potlatch by clans in which chiefs confront one another (Tsimshian); (4) a potlatch of chiefs and brotherhoods (Kwakiutl). But it would take too long to proceed thus. More-over, the differences between three of the four forms (the Tsim-shian form is not dealt with) have been expounded by Davy.[155] Finally, as far as our study is concerned – that of the three themes of the gift, the obligation to give, the obligation to receive and reciprocate – the four forms of the potlatch are comparatively identical.

The three obligations: to give, to receive, to reciprocate

The obligation to give is the essence of the potlatch. A chief must give potlatches for himself, his son, his son-in-law, or his daughter,[156] and for his dead.[157] He can only preserve his authority over his tribe and village, and even over his family, he can only maintain his rank among the chiefs[158] – both nationally and internationally – if he can prove he is haunted and favoured both by the spirits and by good fortune[159], that he is possessed, and also possesses it.[160] And he can only prove this good fortune by spending it and sharing it out, humiliating others by placing them 'in the shadow of his name'.[161] Each Kwakiutl and Haïda noble has exactly the same idea of 'face' as has the Chinese man of letters or officer.[162] It is said of one of the great mythical chiefs who gave no potlatch that he had a 'rotten face'.[163] Here the expression is even more exact than in China. For in the American Northwest, to lose one's prestige is indeed to lose one's soul. It is in fact the 'face' the dancing mask, the right to incarnate a spirit, to wear a coat of arms, a totem, it is really the *persona* – that are all called into question in this way, and that are lost at the potlatch,[164] at the game of gifts,[165] just as they can be lost in war,[166] or through a mistake in ritual.[167] In all such societies one hastens to give. There is not one, single, special moment, even apart from the winter solemnities and gatherings, when one is not obliged to invite one's friends, to share with them the windfall gains of the hunt or food gathering, which come from the gods and the totems.[168] There is not one single moment when you are not obliged to redistribute everything from a potlatch in which you have been the beneficiary;[169] or when you are not obliged to acknowledge by gifts some service or other,[170] whether performed by chiefs,[171] vassals, or relatives:[172] all this under pain of violating etiquette – at least for nobles – and of losing rank.[173]

The obligation to invite is clearly evident when imposed by clans on clans, or tribes on tribes. Indeed it only has meaning if it is offered to others outside the family, the clan, or the phratry.[174] It is essential to invite anyone who can,[175] or wishes to[176] come, or actually turns up[177] at the festival at the potlatch.[178] Failure to do so has fatal consequences.[179] An important Tsimshian myth[180] reveals the mentality in which this essential theme of European folklore also originated: that of the wicked fairy who was forgotten at a baptism or a marriage. The tissue of institutions from which the theme is woven is clearly apparent here. We see in what civilizations it functioned. A princess of one of the Tsimshian villages has conceived in the 'land of the otters' and miraculously gives birth to 'the Little Otter'. She returns with her child to the village where her father is the chief. Little Otter catches a large halibut on which his grandfather regales all his fellow chiefs, from all the tribes. He presents his grandson to everybody and enjoins them not to kill him if they come across him in his animal form while out fishing: 'Here is my grandson who has brought this food for you and which I have served to you, my guests.' In this way the grandfather grew rich with all kinds of goods given him when they came to his home to partake of the whales, seals, and fresh fish that Little Otter brought back during the winter famines. But they had forgotten to invite one chief. So, one day when the crew of a boat belonging to the neglected tribe met Little Otter out at sea, holding a large seal in his mouth, the bowman in the boat killed him and took the seal from him. The grandfather and the other tribes searched for Little Otter until they became aware of what had happened to the forgotten tribe. The latter presented its excuses: it did not know who Little Otter was. His mother, the princess, died of grief. The chief who had been the unwitting culprit brought to the chief, the grandfather, all kinds of gifts to expiate his mistake. And the myth concludes:[181] 'This is why peoples mounted a great festival when the son of a chief was born and was given a name, so that no one should not know who he was.' The potlatch, the distribution of goods, is the basic act of 'recognition', military, juridical, economic, and religious in every sense of the word. One 'recognizes' the chief or his son and becomes 'grateful' to him.[182]

The ritual followed in the Kwakiutl festivals[183] and those of other tribes in this group sometimes expresses this principle of obligatory invitation. It can happen that a part of the festival begins with the Ceremony of the Dogs. The latter are represented by masked men who leave one house and force an entrance into another. This commemorates the time when the people of the three other clans of the Kwakiutl tribe proper omitted to invite the most high-ranking clan among them, the Guetala.[184] The latter did not wish to remain 'profane' and, entering the house where dances were going on, destroyed everything.

The obligation to accept is no less constraining. One has no right to refuse a gift, or to refuse to attend the potlatch.[185] To act in this way is to show that one is afraid of having to reciprocate, to fear being 'flattened' [i.e. losing one's name] until one has reciprocated. In reality this is already to be 'flattened'. It is to 'lose the weight' attached to one's name.[186] It is either to admit oneself beaten in advance[187] or, on the contrary, in certain cases, to proclaim oneself the victor and invincible.[188] Indeed it seems, at least among the Kwakiutl, that an acknowledged position in the hierarchy, and victories in previous potlatches, allow one to refuse an invitation, or even, when present at a potlatch, to refuse a gift without war ensuing. Yet then the potlatch becomes obligatory for the one who has refused; in particular, he must make even richer the 'festival of fat' where this ritual of refusal can in fact be observed.[189] The chief who believes himself to be superior spurns the spoonful of fat presented to him; he goes to fetch his 'copper object' and returns with it in order to 'put out the fire' (of the fat). There follows a succession of formalities that signify the challenge that binds the chief who has refused to give another potlatch, another 'festival of fat'.[190] But in principle every gift is always accepted and even praised.[191] One must voice one's appreciation of the food that has been prepared for one.[192] But, by accepting it one knows that one is committing onself.[193] A gift is received 'with a burden attached'.[194] One does more than derive benefit from a thing or a festival: one has accepted a challenge, and has been able to do so because of being certain to be able to reciprocate,[195] to prove one is not unequal.[196] By confronting one another in this way chiefs can place themselves in a comic situation that is surely perceived as such. As in ancient Gaul or Germany, or at our own banquets for students, soldiers, and peasants, one is committed to gulping down large quantities of food, in order to 'do honour', in a somewhat grotesque way, to one's host. Even if one is only the heir of the person who has made the challenge,[197] it is taken up. To refrain from giving, just as to refrain from accepting,[198] is to lose rank – as is refraining from reciprocating.[199]

The obligation to reciprocate[200] constitutes the essence of the potlatch, in so far as it does not consist of pure destruction. These acts of destruction are very often sacrificial, and beneficial to the spirits. It would seem they need not all be reciprocated unconditionally, particularly when they are the work of a chief recognized in the clan as being superior, or that of a chief of a clan that has itself already been recognized as superior.[201] However, normally, the potlatch must be reciprocated with interest, as must indeed every gift. The rate of interest generally ranges from 30–100 per cent a year. Even if a subject receives a blanket from his chief for some service he has rendered, he will give two in return on the occasion of a marriage in the chief's family, or the enthronement of the chief's son, etc. It is true that the latter, in his turn, will give away all the goods that he obtains at future potlatches, when the opposing clans will heap benefits upon him.

The obligation to reciprocate worthily is imperative.[202] One loses face for ever if one does not reciprocate, or if one does not carry out destruction of equivalent value.[203]

The punishment for failure to reciprocate is slavery for debt. At least, this functions among the Kwakiutl, the Haïda, and the Tsimshian. It is an institution really comparable in nature and function to the Roman *nexum*. The individual unable to repay the loan or reciprocate the potlatch loses his rank and even his status as a free man. Among the Kwakiutl, when an individual whose credit is poor borrows, he is said to 'sell a slave'. There is no need to point out the identical nature of this and the Roman expression.[204]

The Haïda[205] even say – as if they had discovered the Latin expression independently – regarding a mother who gives a present to the mother of a young chief for a betrothal contracted as a minor, that she has: 'put a thread around him'.

But, just as the Trobriand *kula* is only an extreme case of the exchange of gifts, so the potlatch in societies living on the Northwest American coast is only a kind of monstrous product of the system of presents. At least in lands such as those of the Haïda and the Tlingit, where phratries exist, there still remain considerable traces of the onetime 'total services', which, moreover, are so characteristic of the Athapascans, the important group of related tribes. Presents are exchanged for any and every reason, for every 'service', and everything is given back later, or even at once, and is immediately given out again.[206] The Tsimshian are not very far from having kept to the same rules.[207] In numerous cases the rules even appertain apart from the potlatch, among the Kwakiutl.[208] We shall not labour this point, which is self-evident. Older writers describe the potlatch no differently, so much so that one may wonder whether it constitutes a distinct institution.[209] We recall that among the Chinook, one of the least well-known tribes, but which might have been among the most important ones to study, the word potlatch signifies gift.[210]

The force of things

One can push the analysis further and demonstrate that in the things exchanged during the potlatch, a power is present that forces gifts to be passed around, to be given, and returned.

First, at least among the Kwakiutl and Tsimshian, the same distinction is made between the various kinds of property as made by the Romans, the Trobriand peoples, and the Samoans. For these there exist, on the one hand, the objects of consumption and for common sharing[211] (I have found no trace of exchanges). And on the other hand, there are the precious things belonging to the family,[212] the various talismans, emblazoned copper objects, blankets made of skins, or cloth bedecked with emblems. This latter type of object is passed on as solemnly as women hand over at marriage the 'privileges' to their sons-in-law,[213] and names and ranks to children and sons-in-law. It is even incorrect to speak in their case of transfer. They are loans rather than sales or true abandonment of possession. Among the Kwakiutl a certain number of objects, although they appear at the potlatch, cannot be disposed of. In reality these pieces of 'property' are *sacra* that a family divests itself of only with great reluctance, and sometimes never.

More detailed observation among the Haïda will reveal the same distinctions between things. The latter have in fact even made the notion of property and fortune divine, as did the Ancients. Through a mythological and religious effort, fairly infrequent in America, they have raised themselves to a level where they have personified an abstraction. English writers refer to the 'Property Woman' about whom there are myths and of whom we have descriptions.[214] For the Haïda she is nothing less than the mother, the originating goddess of the dominant phratry, that of the Eagles. Yet on the other hand – and this is a strange fact that awakens very distant recollections of the Asiatic and Ancient world – she seems identical to the 'queen',[215] the main protagonist in 'the game of sticks' ('tip-it'), the one who wins everything and whose name she bears in part. This goddess is to be found in the Tlingit lands,[216] and the myth about her, if not the worship of her, among the Tsimshian[217] and the Kwakiutl.[218]

The sum total of these precious things constitutes the magical dower; this is often identical for both donor and *recipient*, and also for the spirit who has provided the clan with these talismans, or the hero who is the originator of the clan and to whom the spirit has given them.[219] In any case all these things are always, and in every tribe, spiritual in origin and of a spiritual nature.[220] More-over, they are contained in a box, or rather in a large emblazoned case[221] that is itself endowed with a powerful personality,[222] that can talk, that clings to its owner, that holds his soul, etc.[223]

Each of these precious things, these signs of wealth possesses – as in the Trobriand Islands – its individuality, its name,[224] its qualities, its power.[225] The large abalone shells,[226] the shields that are covered with these shells, the belts and blankets that are decorated with them, the blankets themselves[227] that also bear emblems, covered with faces, eyes, and animal and human figures that are woven and embroidered on them – all are living beings. The houses, the beams, and the decorated walls[228] are also beings. Everything speaks – the roof, the fire, the carvings, the paintings – for the magical house is built,[229] not only by the chief or his people, or the people of the opposing phratry, but also by one's gods and ancestors. It is the house that both accepts and rejects the spirits and the youthful initiates.

Each one of these precious things[230] possesses, moreover, pro-ductive power itself.[231] It is not a mere sign and pledge; it is also a sign and a pledge of wealth, the magical and religious symbol of rank and plenty.[232] The dishes[233] and spoons[234] used solemnly for eating, and decorated, carved, and emblazoned with the clan's totem or the totem of rank, are animate things. They are replicas of the inexhaustible instruments, the creators of food, that the spirits gave to one's ancestors. They are themselves deemed to have fairylike qualities. Thus things are mixed up with spirits, their originators, and eating instruments with food. The dishes of the Kwakiutl and the spoons of the Haïda are essential items that circulate according to very strict rules and are meticulously shared out among the clans and the families of the chiefs.[235]

The 'money of fame'[236]

Yet above all it is the emblazoned copper objects[237] that, as basic goods for the potlatch, are the focus of important beliefs and even of a cult.[238] First, in every tribe there exists a cult and a myth regarding copper,[239] which is regarded as a living thing. Copper, at least among the Haïda and the Kwakiutl, is identified with the salmon, which is itself the object of a cult.[240] Yet, besides this element of metaphysical and technical mythology,[241] all these pieces of copper are, each one separately, the subject of individual and particular beliefs. Each important piece of copper in the families of the clan chiefs has its name,[242] its own individuality, its own value,[243] in the full sense of the word – magical, economic, permanent, and perpetual – despite the vicissitudes of the potlatch through which they pass, and even beyond the partial or complete acts of destruction they suffer.[244]

Moreover, they have a power of attraction that is felt by other copper objects, just as wealth attracts wealth, or dignities bring honours in their train, as well as the possession of spirits and fruitful alliances,[245] – and *vice versa*. They are alive and move autonomously,[246] and inspire other copper objects to do so.[247] One of them[248] is called among the Kwakiutl 'the attracter of copper objects', and the story depicts how the copper objects group around it. At the same time the name of its owner is 'property that flows towards me'. Another frequent name for copper objects is: 'the bringer of property'. Among the Haïda and the Tlingit the copper objects form a 'strongpoint' around the princess who brings them;[249] elsewhere the chief who has them in his possession[250] is rendered invincible. They are 'the flat, divine things'[251] of the household. Often the myth identifies them all, the spirits that have given the copper objects,[252] their owners, and the copper objects themselves.[253] It is impossible to distinguish what makes the strength of spirit in the one and wealth in the other: the copper object speaks, and grumbles.[254] It demands to be given away, to be destroyed; it is covered with blankets to keep it warm, just as the chief is buried under the blankets that he is to share out.[255]

Yet, on the other hand, at the same time as goods, it is wealth[256] and good luck that are passed on. It is the initiate's spirit, it is his attendant spirits, that give the initiate possession of copper objects, of talismans that themselves are the means of acquiring other things: other copper objects, wealth, rank, and finally spirits, all things that, moreover, are of equivalent value. All in all, when one considers both the copper objects and the other permanent forms of wealth that are likewise an object of hoarding and of alternating potlatches, masks, talismans, etc. – all are mingled together as regards use and effect.[257] Through them one obtains rank; it is because one obtains wealth that one obtains a spirit. The latter, in its turn, takes possession of the hero who has overcome all obstacles. Then again, this hero has his Shaman trances, his ritual dances, and the services of his government [sic] paid for him. Everything holds together, everything is mixed up together. Things possess a personality, and the personalities are in some way the permanent things of the clan. Titles, talismans, copper objects, and the spirits of the chiefs are both homonyms and synonyms[258] of the same nature and performing the same function. The circulation of goods follows that of men, women, and children, of feasts, rituals, ceremonies, and dances, and even that of jokes and insults. All in all, it is one and the same. If one gives things and returns them, it is because one is giving and returning 'respects' – we still say 'courtesies'. Yet it is also because by giving one is giving *oneself*, and if one gives *oneself*, it is because one 'owes' *oneself* – one's person and one's goods – to others.

First conclusion

Thus, in four important population groups we have discovered the following: first, in two or three groups, the potlatch; then the main reason for, and the normal form of the potlatch itself; and what is more, beyond the potlatch, and in all these groups, the archaic form of exchange – that of gifts presented and reciprocated. Moreover, we have identified the circulation of things in these societies with the circulation of rights and persons. We could, if we wanted, stop there. The number, extent, and importance of these facts justifies fully our conception of a regime that must have been shared by a very large part of humanity during a very long transitional phase, one that, moreover, still subsists among the peoples we have described. These phenomena allow us to think that this principle of the exchange-gift must have been that of societies that have gone beyond the phase of 'total services' (from clan to clan, and from family to family) but have not yet reached that of purely individual contract, of the market where money circulates, of sale proper, and above all of the notion of price reckoned in coinage weighed and stamped with its value.

3

SURVIVALS OF THESE PRINCIPLES IN ANCIENT SYSTEMS OF LAW AND ANCIENT ECONOMIES

All the facts set out have been gathered in what we term the ethnographical field. Moreover, they concentrate on the societies that people the borders of the Pacific.[1] Usually these kinds of facts are used as curiosities, or at the most for comparison, to gauge how far our own societies are removed from or resemble these kinds of institutions, which are termed 'primitive'.

However, they have general sociological value, since they allow us to understand a stage in social evolution. But there is more to them than this: they have also a bearing on social history. Institutions of this type have really provided the transition towards our own forms of law and economy. They can serve to explain historically our own societies. The morality and the practice of exchanges employed in societies immediately preceding our own still retain more-or-less important traces of all the principles we have just analysed. We believe, in fact, that we are in a position to show that our own systems of law and economies have emerged from institutions similar to those we describe.[2]

We live in societies that draw a strict distinction (the contrast is now criticized by jurists themselves) between real rights and personal rights, things and persons. Such a separation is basic: it constitutes the essential condition for a part of our system of property, transfer, and exchange. Now, this is foreign to the system of law we have been studying. Likewise our civilizations, ever since the Semitic, Greek, and Roman civilizations, draw a strong distinction between obligations and services that are not given free, on the one hand, and gifts, on the other. Yet are not such distinctions fairly recent in the legal systems of our great civilizations? Have these not gone through a previous phase in which they did not display such a cold, calculating mentality? Have they not in fact practised these customs of the gift that is exchanged, in which persons and things merge? The analysis of a few features of Indo-European legal systems will allow us to demonstrate that they have, indeed, undergone this meta-morphosis. In Rome we shall find traces of this. In India and ancient Germany it will be the laws themselves, still very much alive, that we shall still see functioning in a comparatively recent era.

I
PERSONAL LAW AND REAL LAW (VERY ANCIENT ROMAN LAW)

A comparison between these archaic laws and Roman law that predates the era, somewhat earlier than before it really becomes historical,[3] and Germanic law at the time when it does likewise,[4] throws light on both types of law. In particular, it allows us to pose once again one of the most controversial questions in the history of law, the theory of the *nexum*.[5]

In a study that has done more than merely shed light on the matter,[6] Huvelin has compared the *nexum* to the Germanic *wadium* and generally to the 'additional pledges' (Togo, the Caucasus, etc.) given upon the occasion of a contract and has then compared the pledges to the sympathetic magic and power that is given to the other contracting party by anything that has been in contact with the party proposing the contract. But such an explanation is valid for only some of the facts. The sanction of magic is only *possible*, and is itself only the consequence of the nature and spiritual character of the thing that is given. First, the additional pledge, and in particular the Germanic *wadium*,[7] are more than the exchange of pledges, even more than life pledges designed to establish a possible magical hold over the other party. The thing pledged is normally without value: for example, sticks are exchanged, the *stips* in the 'stipulation' of Roman law[8] and the *festuca notata* in the German 'stipulation'; even the pledges given on account,[9] of Semitic origin, are more than advances. They are things that themselves are animate. Above all, they are still the residues of former obligatory gifts, that were owed because of reciprocity. The contracting parties are bound by them. In this respect these additional exchanges express, as a fiction, that coming and going of souls and things that are all intermingled with one another.[10] The *nexum*, the legal 'lien', springs from things as much as from men.

This very formalism proves the importance of things. In Quiritary Roman law, the hand over of property – and essential items of property were slaves and cattle, and later immovable property – was in no way common, profane, or simple. The handing over is always solemn and reciprocal.[11] It is still carried out as a group: the five witnesses, who at least were friends, and the 'weigher' [arbitrator]. It is mixed up with all kinds of considerations that are foreign to our purely modern juridical and economic conceptions. The *nexum* that is established is thus, as Huvelin clearly perceived, replete with those religious representations that he merely considered as relating too exclusively to magic.

It is true that the *nexum*, the most ancient form of contract in Roman law, is already separated from the substance of collective contracts and also from the ancient system of gifts that commit one. The prehistory of the Roman system of obligations can never be written with certainty. However, we believe we can point to how it might be investigated.

There is certainly a tie expressed by things, *in addition to* magical and religious ties: those of the words and actions of juridical formalism.

This link is still marked by a few very ancient legal terms in the law of the Latins and the peoples of ancient Italy. The etymology of a certain number of these terms seems to point in this direction. We set out the following as a hypothesis.

Originally – so much is sure – things themselves had a personality and an inherent power. Things are not the inert objects that the law of Justinian and our own legal systems conceive them to be. First, they form part of the family: the Roman *familia* includes the *res*, and not only people. We still have a definition of them in the *Digesta*.[12] It is most remarkable that, the farther one goes back in Antiquity the more the meaning of the word *familia* denotes the *res* that are part of it, even going so far as to include food and the family's means of subsistence.[13] The best etymology of the word *familia* is without doubt that which compares it[14] to the Sanskrit *dhaman*, 'house'.

Moreover, things were of two kinds. A distinction was made between *familia* and *pecunia*, between the things of the household (slaves, horses, mules, donkeys) and the cattle subsisting in the fields, far from the stables.[15] A distinction was also made between the *res mancipi* and the *res nec mancipi*, according to the forms of sale.[16] As regards the former, which are made up of precious things, including immovable goods and even children, no disposal of them could take place save according to the precepts of the *mancipatio*,[17] the 'taking (*capere*) in hand (*manu*)'. There is much discussion as to whether the distinction between *familia* and *pecunia* coincided with that between *res mancipi* and *res nec mancipi*. For us there is not a shadow of doubt that this coincidence originally existed. The things that did not fall under the *mancipatio* are precisely the small livestock in the fields and the *pecunia*, money, the idea, word, and form of which derived from cattle and sheep. It might be said that the Roman *veteres* make the same distinction that we have just noted in the areas where the Tsimshian and the Kwakiutl live, between the permanent and essential goods of the 'house' (as they still say in Italy and in France) and the things that are transitory: food, the cattle on the distant pastures, metals, and money, in which, after all, even the non-'emancipated' sons could trade.

Next, originally the *res* need not have been the crude, merely tangible thing, the simple, passive object of transaction that it has become. It would seem that the best etymology is one that compares the word to the Sanskrit *rah, ratih*,[18] gift, present, something pleasurable. The *res* must above all have been something that gives pleasure to another person.[19] Moreover, the thing is always stamped by a seal, as a mark of family property. Thus one can understand that a solemn handing over,[20] *mancipatio*, creates for these things, the *mancipi*, a legal tie. For, in the hands of the *accipiens*, the thing handed over continues, in part and for a time, to belong to the 'family' of the original owner. It remains bound to him, and binds its present possessor until the latter is freed by the execution of the contract, namely by the compensatory handing over of the thing, price, or service that, in turn, will bind the initial contracting party.

SCHOLIUM – EXPLANATORY NOTE

The notion of the power inherent in a thing has, moreover, in two aspects, never been absent from Roman law: theft, *furtum*, and contracts, *re*.

As regards theft,[21] the actions and obligations that it entails are patently due to the power inherent in the thing. It possesses within it an *aeterna auctoritas*,[22] which makes itself felt when it is stolen and lost for good. In this respect the Roman *res* does not differ from Hindu or Haïda property.[23]

The *re* contracts constitute four of the most important legal contracts: borrowing, deposit, pledge, and *commodate*. A certain number of *innominate* contracts also – particularly those we believe to have been, with the contract of sale, at the origin of contract itself: gift and exchange[24] – are likewise said to be *re* ones. But this was inevitable. Indeed, even in our present legal systems, as in Roman law, here it is not possible to circumvent[25] the most ancient rules of law: there must be a thing or service for there to be a gift, and the thing or service must place one under an obligation. For example, it is evident that the cancelling of a gift made on grounds of ingratitude, which occurs in late Roman law,[26] but which is a constant factor in our legal systems, is a normal, perhaps even natural, legal institution.

But these isolated facts are proof only in the case of certain contracts. Our thesis is a more general one. We believe that there cannot have been, in the very distant eras of Roman law, a single instant in which the act of the *traditio* of a *res* was not – even in addition to words and documents – one of the essential elements. Roman law, moreover, has always vacillated on this question.[27] If, on the one hand, it declares that the solemnity of the exchanges, and at least the contract, is necessary, as the archaic laws we have described lay down, if it said: *nunquam nuda traditio transfert dominium*,[28] it likewise declared, although in an era as late as that Diocletian (A.D. 298):[29]*Traditionibus et usucapionibus dominia, non pactis transferuntur.* The *res*, a service or thing, is an essential element in the contract.

Furthermore, all these questions, which have been much debated, are problems that relate to vocabulary and concepts, and, in view of the poverty of the ancient sources, ones that we are very badly qualified to resolve.

Up to this point we have been very sure of our facts. However, it may be permissible to go even farther and perhaps point out to jurists and linguists a broad highway along which search can proceed. At the end of this one may perhaps envisage that a whole legal system had already collapsed by the time of the Twelve Tables, and perhaps well before then. Legal terms other than *familia* and *res* lend themselves to an in-depth study. We shall sketch out a series of hypotheses, each one of which singly is perhaps not very important, but all of which as a whole cannot fail to constitute a fairly considerable body of law.

Almost all the terms of contract and obligation, and a certain number of the forms these contracts take, seem to link up with that system of spiritual bonds created through the crude fact of *traditio*.

First, the contracting party is *reus*,[30] he is above all the person who has received the *res* of another, and thereby becomes his *reus*, i.e. the individual who is linked to him by the thing itself, namely by his spirit.[31] The etymology of this has already been suggested. It has often been contradicted on the grounds that it made no sense. On the contrary, its meaning is very clear. Indeed, as Hirn notes,[32] *reus* is originally a genitive in *-os* of *res* and replaces *rei-jos*. It is the man possessed by the thing. It is true that Hirn, and Walde who reproduces it,[33] here translate[34] *res* by 'trial' and *rei-jos* by 'involved in the trial'. But this translation is arbitrary, and assumes that *res* is above all a procedural term. On the contrary, if our semantic derivation is accepted, since every *res* and every *traditio* of *res* is the object of an 'affair', a public 'trial', it can be realized that the meaning of being 'involved in the trial' is, on the contrary, a secondary meaning. It is even more the case that the meaning of 'guilty' for *reus* is one still more derived. We would trace the genealogy of the meaning in a way that is directly the opposite to that ordinarily followed. We would say: (1) the individual possessed by the thing; (2) the individual involved in the matter caused by the *traditio* of the thing; (3) finally, the guilty one and the one responsible.[35] From this viewpoint all the theories of 'quasi-offence' as the origin of a contract, of *nexum* and *actio*, become a little clearer. The mere fact of having the thing puts the *accipiens* in an uncertain state of quasi-culpability (*damnatus, nexus, aere obaeratus*), of spiritual inferiority and moral inequality (*magister, minister*),[36] in relation to the one delivering (*tradens*) the contract.

We likewise link to this system of ideas a certain number of very ancient characteristics of the form still practised, if not understood, of the *mancipatio*,[37] of 'buying and selling', which was to become the *emptio venditio*[38] in very ancient Roman law. In the first place we must note that it always includes a *traditio*.[39] The first possessor, the *tradens*, displays his property, solemnly detaches himself from his thing, hands it over, and thus 'buys' the *accipiens*. In the second place, the *mancipatio* proper corresponds to this operation. He who receives the thing takes it in his *manus*, and not only acknowledges it as accepted, but acknowledges that he himself is 'sold' until payment has been made. Following the prudent Romans, we customarily consider only the *mancipatio* and understand it as only the assumption of possession, but there are several symmetrically corresponding acts of taking possession, of things and persons, involved in the same operation.[40]

Moreover, there has been much lengthy discussion as to whether the *emptio venditio*[41] corresponds to two separate acts or a single one. As can be seen, we provide yet another reason for saying that we must reckon with two, although they can follow each other almost immediately in the case of a cash sale. Just as there is in more primitive forms of law the gift, and then the gift reciprocated, in ancient Roman law there is the putting-up for sale, and then the payment. Under these conditions one has no difficulty in understanding the whole system, and, in addition, even the act of stipulation.[42]

It is in fact almost enough to note the solemn formulas used: that of the *mancipatio*, concerning the bronze ingot, that of the acceptance of the gold of the slave buying his freedom[43] (this gold 'must be pure, of good quality, profane, and his own' – '*puri, probi, profani, sui*'). The formulas are identical. Moreover they are both echoes of the formulas employed in the oldest form of *emptio*, that of cattle and the slave, which have been preserved for us in the form of *jus civile*.[44] The second possessor only accepts the thing if it is free from defects, and above all, from defects of magic. And he only accepts it because he can reciprocate or compensate, or pay the price. One should note the expression: '*reddit pretium, reddere*', etc. in which there still appears the root *dare*.[45]

Moreover, Festus has clearly preserved for us the meaning of the term *emere* (to buy), and even the form of law that it expresses. He also says: '*abemito significat demito vel auferto; emere enim antiqui dicebant pro accipere*' (see under *abemito*) and, moreover, he comes back to this meaning: '*Emere quod nunc est mercari antiqui accipiebant pro sumere*' (see under *emere*), which, moreover, is the meaning of the Indo-European word to which the Latin word itself relates. *Emere* is to take, to accept something from someone.[46]

The other term of *emptio venditio* likewise seems to sound a juridical note different from that of the prudent Romans,[47] for whom there was only barter and gift when there was neither price nor money, the signs of sale. *Vendere*, originally *venum-dare*, is a composite word of an archaic prehistoric type.[48] Without any doubt it clearly includes an element *dare*, which reminds us of both gift and transfer. For the other element, it clearly seems to borrow an Indo-European term that already meant, not sale, but the price of sale, ώνή, Sanskrit *vasnah*, which Hirn[49] has moreover compared to the Bulgarian word that means dowry, the buying price of a wife.

OTHER INDO-EUROPEAN LAW SYSTEMS

These hypotheses concerning very ancient Roman law relate somewhat to a prehistoric order. The law, morality, and economy of the Latins must have had these forms, but they were forgotten when their institutions entered the historical period. For it is precisely the Romans and Greeks,[50] who, perhaps, following upon the Semites of the north and west,[51] invented the distinction between personal and real law, separated sale from gift and exchange, isolated the moral obligation and contract, and in particular, conceived the difference that exists between rites, laws, and interests. It was they who, after a veritable, great, and admirable revolution, went beyond all the outmoded morality, and this economy of the gift. It was too dependent on chance, was overexpensive and too sumptuous, burdened with consideration for people, incompatible with the development of the market, commerce, and production, and, all in all, at that time was anti-economic.

Furthermore, all our work of reconstitution leads only to a likely hypothesis. Yet the degree of probability in any case increases, because other Indo-European systems of law, authentic written law, have undoubtedly known, in times that are already historical and relatively close to our own, a system of the kind that we have described in the societies of Oceania and America, which are commonly called primitive, and are, at the very most, archaic. Accordingly, we can generalize with some certainty.

The two kinds of Indo-Germanic law that have best preserved

traces of this are Germanic and Hindu law. They are also ones for which we have numerous written texts.

II
CLASSICAL HINDU LAW[52]: THEORY OF THE GIFT

N.B. In using Hindu juridical documents one encounters a fairly serious difficulty. The codes and the epic books that have as much authority were drawn up by the Brahmins, and, one may say, if not for them, at least for their advantage, in the very era of their triumph.[53] They show us only a theoretical law system. Thus it is only by an effort at reconstitution, with the help of the numerous admissions that they contain, that we can catch a brief glimpse of what kind of law and economy prevailed among the two other castes, the *ksatriya* and *vaicya*. In the event, the theory, the 'law of the gift' that we are about to describe, the *danadharma*, is only really applicable to the Brahmins, to the way they solicit and receive a gift – without reciprocating it save through their religious services – and also to the way a gift is their due. Naturally it is this duty of giving to the Brahmins that is subject to numerous injunctions. It is probable that quite different relationships prevailed among the nobility, the princely, families, and, within the numerous castes and races, among the common people. We can scarcely surmise what they were. It is not important: the facts regarding Hindu practice are very widely known.

Ancient India immediately after the Aryan colonization was in fact a land of the potlatch twice over.[54] First, the potlatch is still found among two very large groups that were once much more numerous, forming the substratum of a great part of the Indian population: the tribes of Assam (Tibeto-Burman) and the tribes of *munda* origin (Austro-Asiatic). We are even entitled to suppose that the tradition of these tribes is the one that subsisted, but against a Brahmin background.[55] For example, one might see the vestiges[56] of an institution comparable to that of the Batak *indjok* and to the other principles of Malayan hospitality in rules that forbid one to eat without having invited the unexpected guest to do so as well: 'he eats the *halahalah* poison, [he who eats] without his friend joining in'. Moreover, institutions of the same kind, if not of the same species, have left a few traces in the most ancient Veda. As they are found over almost all the Indo-European world,[57] we have reason to believe that they were brought by the Aryans to India.[58] The two tendencies doubtless merged at a time that can almost be fixed as contemporaneous with the latter parts of the Veda and with the colonization of the two great plains of the two great rivers, the Indus and the Ganges. Undoubtedly these two traditions also reinforced one another. Thus, as soon as we leave behind the Vedic era of literature, we find that this theory has been extraordinarily developed, as have its usages. The Mahabharata is the story of a gigantic potlatch: the game of dice of the Kauravas against the Pandavas; jousting tournaments and the choice of bridegrooms by Draupadi, the sister and polyandrous wife of the Pandavas.[59] Other repetitions of the same legendary cycle are to be met with in the finest episodes of the epic – for example, the romance of Nala and Damayanti, as does the whole Mahabharata, tells of the construction and assembling of a house, a game of dice, etc.[60] But everything is distorted by the literary and theological flavour of the story.

However, the argument we are setting out does not necessitate our analysing these diverse origins and building a hypothetical reconstruction of the entire system.[61] Likewise, the number of classes that were concerned, and the time when it flourished, need not be elaborated in detail in a comparative study. Later, for reasons that do not concern us here, this system of law disappears, except as it works in favour of the Brahmins. Yet one may say that it was certainly in force for six to ten centuries, from the eighth century B.C. to the second or third century A.D. This is sufficient: the epic and the Brahmin law still survive in the old atmosphere: presents are still obligatory, things have special powers and form part of human persons. Let us confine ourselves to describing these forms of social life and studying the reasons for them. A simple description will be fairly conclusive.

The thing that is given produces its rewards in this life and the next. Here in this life, it automatically engenders for the giver the same thing as itself:[62] it is not lost, it reproduces itself; in the next life, one finds the same thing, only it has increased. Food given is the food that in this world will return to the giver; it is food, the same food, that he will find in the other world. And it is still food, the same food, that he will find in the series of his reincarnations.[63] The water, wells, and springs that are given ensure against thirst;[64] the clothes, gold, and sunshades, the sandals that allow one to walk on the scorching-hot ground, come back to the giver both in this life and the next. The land given away that yields its harvests for others, causes the affairs of the donor to prosper both in this world and in the next as well as in future rebirths. 'As the waxing of the moon increases day by day, likewise the gift of land once made grows from year to year (from one harvest to the next).'[65] The earth produces its harvests, its income and taxes, mines and cattle. The gift of it once made enriches both giver and recipient with these same products.[66] All such juridico-economic theology is developed in infinitely magnificent phrases, in innumerable *centos* of verse, and neither legal codes nor epics cease to harp upon this theme.[67]

The land, the food, and all that one gives are, moreover, personified: they are living creatures with whom one enters into a dialogue, and who share in the contract. They seek to be given away. The land once spoke to the sun hero, Rama, the son of Jamadagni, and when the latter heard its song, he gave it away in its entirety to king Kacyapa himself. The land[68] told him in its own, doubtless ancient language:

(margin note: karma & the three-fold law)

Receive me (the recipient);
Give me (the donor);
By giving you will obtain me once more.

[handwritten margin note: Christianity — God gave Christ to people; Christ sacrificed himself for people; people got Heaven; God got his son back]

And it added, speaking this time in a somewhat Low Brahmin tongue: 'In this world and in the next, what is given away is acquired once more.' A very old code[69] tells how Anna (food itself made into a god), declaimed the following verse:

> He who, without giving me to the gods, to the shades, to his servants and guests, consumes me when prepared, and in his madness (thus) swallows poison, I consume him, I am his death.
> But for him who offers up the *agnihotra*, accomplishes the *vaiçvadeva*,[70] and then eats – contentedly, in purity and faith – what remains after he has fed those that he should feed, for him I become ambrosia, and he has pleasure in me.

It is in the nature of food to be shared out. Not to share it with others is 'to kill its essence', it is to destroy it both for oneself and for others. This is the interpretation, both materialist and idealist, that Brahminism has given to charity and hospitality.[71] Wealth is made to be given away. If there were no Brahmins to receive it, 'vain would be the riches of the rich'.[72] 'He who eats without knowledge kills the food, and once it is eaten, it kills him.'[73] Avarice breaks the circle of the law; rewards and foods are perpetually reborn from one another.[74]

Moreover Brahminism, in this interplay of exchange, as well as in theft, has clearly identified property with the person. The property of the Brahmin is the Brahmin himself. 'The Brahmin's cow is a poison, a poisonous snake', states already the Veda of the magicians.[75] The old code of Baudhayana[76] declares: 'The property of the Brahmin kills [the guilty one] with his sons and grandsons; poison is not [poison]; the property of the Brahmin is called poison [par excellence].' Property contains in itself its own sanction because it represents what is terrible in the Brahmin. There is even no need for the theft of the Brahmin's property to be conscious and intended. The entire 'reading' of our Parvan,[77] of the section of the Mahabharata that most concerns us, tells how Nrga, the king of the Yadus, was changed into a lizard, through the sin of his own people, for having given a cow to a Brahmin that belonged to another Brahmin. He who had accepted it in good faith did not want to give it back, not even in exchange for a 100,000 others. It was a part of his household, it constituted one of his own:

> It is adapted to the place and the times, it is a good milker, peaceable and very devoted. Its milk is sweet, it is a precious and permanent good in my household. (line 3466)
> It (this cow) nourishes a little child of mine who is weak and has been weaned. It cannot be given away by me. (line 3467)

Nor will the one from whom the cow has been removed accept another. It is irrevocably the property of both Brahmins. Faced with the two refusals, the unfortunate king remains under a spell for thousands of years because of the curse that the property carried with it.[78]

Nowhere is the link between the thing given and the donor, between property and its owner, closer than in the rules concerning the gift of the cow.[79] They are celebrated. By observing them, by feeding himself on barley and cow dung, by sleeping on the ground, king Dharma[80] (the law), Yudhisthira himself, the principal hero of the epic, becomes a 'bull' among kings. For three days and nights the owner of the cow imitates it and. observes 'the wishes of the cow'.[81] He feeds himself solely on the 'juices of the cow' – water, dung, and urine – one night out of three. (In the urine resides Çri (Fortune) herself.) One night out of three he lies down with the cows on the ground, as they do, and, adds the commentator, 'without scratching himself, without disturbing the vermin', thus identifying himself, 'in one single soul, with them'.[82] When he enters the stable, calling the cows by sacred names,[83] he adds 'the cow is my mother, the bull is my father', etc. He will repeat the above formula during the act of giving. There comes the solemn moment of transferral. After singing the praises of cows the recipient says:

> Those which you are, those I am, become this day of your essence. By giving you away, I gave myself.[84] (verse 3676)

and receiving the gift, the recipient (performing the *pratigra-hana*),[85] says:

> Mutated (transmitted) in spirit, received in the spirit, let us both glorify each other, you in the forms of the Soma (moonlike) and Ugra (sunlike).[86] (verse 3677)

Other principles of Brahmin law remind us strongly of certain Polynesian, Melanesian, and American customs that we have described. The way of receiving the gift is curiously similar. The Brahmin has an invincible sense of pride. First, he refuses to have anything to do with the transaction. He must not even accept anything that emerges from it.[87] In a national economy where there were towns, markets, and money, the Brahmin remains true to the type of economy and morality of the ancient Indo-Iranian pastoral peoples, as well as to that of the non-aboriginal and aboriginal peasants of the great plains. He even retains that dignified attitude of the noble,[88] who is the more offended the more one heaps upon him.[89] Two 'readings' in the Mahabharata recount how the seven *rsi*, the great seers, and their retinue, in a time of famine, when they were about to eat the body of the son of the king Cibi, refused the huge presents and even the golden figs that were offered them by king Çaivya Vrsadarbha, and replied to him:

> O king, to receive from kings is at first honey, at the end, poison. (verse 445 and reading 93, verse 34)

There follow two series of curses. The whole theory is even somewhat comical. This entire caste, which lives on gifts, claims to refuse them.[90] Then it gives way and accepts those that have been spontaneously offered.[91] Then it draws up long lists[92] of the people from whom it can accept gifts, in what circumstances, and what kind of things,[93] going so far as to allow everything in the case of famine,[94] subject, it is true, to minor acts of expiation being carried out.[95]

This is because the bond established between donor and recipient is too strong for both of them, as in all the systems we have studied previously, and here still more so, they are too closely linked with one another. The recipient puts himself in a position of dependence vis-à-vis the donor.[96] This is why the Brahmin must not 'accept' gifts, and even less solicit them, from the king. A divinity among divinities, he is superior to the king, and would demean himself if he did anything other than take gifts. Moreover, for the king the way of giving is as important as what he gives.[97]

The gift is therefore at one and the same time what should be done, what should be received, and yet what is dangerous to take. This is because the thing that is given itself forges a bilateral, irrevocable bond, above all when it consists of food. The recipient is dependent upon the anger of the donor,[98] and each is even dependent on the other. Thus one must not eat in the home of one's enemy.[99]

All kinds of archaic precautions are taken. The codes and the epic expatiate upon the theme – as Hindu writers well knew how to do – that gifts, givers, and things given are terms to be considered relatively,[100] after going into details and with scruples, so no error is committed in the way one gives and receives. It is all a matter of etiquette; it is not like in the market where, objectively, and for a price, one takes something. Nothing is unimportant.[101] Contracts, alliances, the passing on of goods, the bonds created by these goods passing between those giving and receiving – this form of economic morality takes account of all this. The nature and intentions of the contracting parties, the nature of the thing given, are all indivisible.[102] The lawyer poet knew how to express perfectly what we seek to describe:

Here there is nothing save a wheel (turning only one way).[103]

III
GERMANIC LAW (THE PLEDGE AND THE GIFT)

Although Germanic societies have not preserved for us such ancient and complete vestiges of their theory of the gift,[104] they nevertheless had a system of exchanges of gifts, given, received, and reciprocated either voluntarily or obligatorily, so clearly defined and well developed that there are few systems so typical.

Germanic civilization was itself a long time without markets.[105] It remained an essentially feudal and peasant society, and the notions and even the terms 'buying price' and 'selling price' seem to be of recent origin.[106] In earlier times it had developed to the extreme the entire system of potlatch, but in particular, the complete system of gifts. The clans within the tribes, the large undivided families within the clans,[107] the tribes one with another, the chiefs among themselves, and even the kings among themselves – all lived to a fairly large extent morally and economically outside the closed confines of the family group. Thus, it was by the form of the gift and the alliance, by pledges and hostages, by feasts and presents that were as generous as possible, that they communicated, helped, and allied themselves to one another. We have seen earlier the whole range of presents borrowed from the Havamal. In addition to this beautiful landscape of the *Edda* saga, we shall point out three facts.

A detailed study of the very rich German vocabulary of the words derived from *geben* and *gaben* has not yet been made.[108] They are extraordinarily numerous: *Ausgabe, Abgabe, Hingabe, Liebesgabe, Morgengabe,* the very curious *Trostgabe* (consolation prize), *vorgeben, vergeben* (to waste, and to forgive), *widergeben* and *wiedergeben.* The study of *Gift, Mitgift,* etc., and the study of the institutions that are designated by these words has also yet to be made.[109] On the other hand, the whole system of presents and gifts, its importance in tradition and folklore, including the obligation to reciprocate, are admirably described by Richard Meyer in one of the most appealing works of folklore that we know.[110] We can merely refer to it, for the time being drawing attention to the well-chosen remarks concerning the strength of the bond that imposes an obligation, the *Angebinde* that makes up exchange, the offer, the acceptance of that offer, and the obligation to reciprocate.

There is, moreover, an institution that only a short time ago persisted, that doubtless still persists in the morality and economic customs of German villages, and that has an extraordinary importance from the economic viewpoint: it is the *Gaben*,[111] the exact equivalent of the Hindu *adanam*. At baptisms, first communions, engagement parties, and weddings, the guests – who often include the whole village – for example, after the wedding breakfast, or on the previous day, or the following day (*Guldentag*), present gifts whose value generally greatly exceeds the expense of the wedding. In certain areas of Germany, the *Gaben* constitutes the bride's dowry, which is given to her on the wedding morning. This is known as the *Morgengabe*. In a few places the generosity of these gifts is proof of the fertility of the young couple.[112] The contract made through an engagement, the various gifts that godfathers and godmothers make at different times in life to assist and help (*Helfete*) their grandchildren, are equally important. We recognize this theme, which is still well known in all our own customs, folk tales, and legends concerning the invitation, the curse of those not invited, and the blessing and generosity of those invited, particularly when they are fairies.

A second institution has the same origin. It is the need for a 'pledge' in all sorts of Germanic contracts.[113] The very word *gage* (Fr.) comes from this, from *wadium* (cf. English, *wage*). Huvelin[114] has already shown that the Germanic *wadium*[115] provided a means of understanding the binding tie of contracts and compared it to the Roman *nexum*. Indeed, as Huvelin interpreted it, the pledge accepted allows the contracting parties in Germanic law to react with one another, since each possesses something of the other. The other, having been the owner of the thing, may have cast a spell upon it, and the pledge was frequently cut in two, half kept by one party and half by the other. Yet upon such an explanation it is possible to superimpose another more accurate one. The magic sanction can intervene. It is not the sole bond. The thing itself, given and committed in the pledge, is a bond by virtue of its own power. First, the pledge is compulsory. In Germanic law any contract, whether for sale or purchase, for loan or deposit, includes the constitution of a pledge. An object, generally of little value, is given to the other contracting party; a glove, a coin (*Treugeld*), a knife – or (as in France today) pins that will be returned when payment for the thing handed over has been made. Huvelin already notes that the thing is of little value, and normally is personal. He rightly compares this act with the theme of the 'life-token'.[116] The thing passed on in this way is indeed very much infused with the individuality of the donor. The fact that it is in the hands of the recipient stimulates the contracting party to carry out the contract, to redeem himself by redeeming the thing. Thus the *nexum* is in the thing, and not merely in the magical acts, or only the solemn forms of the contract, the words, the oaths, the rituals exchanged, or the shaking of hands. It is in it, as it is in the documents, the 'acts' of magical value, and the 'tallies' that each contracting party retains, the meals taken in common, in which everyone partakes of the substance of everybody else.

Two features of the *wadiatio* demonstrate, moreover, this power that is inherent in the thing. First, the pledge is not only a binding obligation, but also binds the honour,[117] authority, and *mana* of the one who hands it over.[118] The latter remains in a position of inferiority so long as he is not freed from his pledge-wager. For the word *Wette, wetten*,[119] that the *wadium* of the laws translates, has as much the meaning of 'wager' as of 'pledge'. It is the prize of a competition and the sanction of a challenge even more directly than it is a means of constraint upon the debtor. So long as the contract is not completed, he is, as it were, the loser of the wager, the one who comes second in the race. Thus he loses more than he commits himself to, more than he will have to pay. This is without taking into account that he runs the risk of losing what he has received, which the owner will claim back from him, so long as the pledge has not been redeemed.

The second characteristic shows the danger inherent in receiving the pledge, for it is not only the giver who commits himself: the recipient also binds himself. Just like the recipient in the Trobriand Islands, he is wary of the thing that has been given. Thus, it is thrown down at his feet[120] when it is a *festuca notata*,[121] ornamented with Runic characters and notches. When it is a tally of which he may keep a part or not, he receives it on the ground or in his lap (*in laisum*), and not in his hand. The whole ritual takes the form of a challenge and is full of mistrust, giving expression to both. Moreover, in English, at one time 'to cast (down) the gage' was equivalent to the phrase 'to throw (down) the gauntlet'. This is because the pledge, the 'gage', like the thing that is given, holds danger for both parties.

There is a third fact. The danger represented by the thing given or handed on is doubtless nowhere better sensed than in the very ancient Germanic law and languages. This explains the double meaning of the word *Gift* in all these languages – on the one hand, a gift, on the other, poison. We have traced elsewhere the semantic history of this word.[122] This theme of the fatal gift, the present or item of property that is changed into poison is fundamental in Germanic folklore. The Rhine gold is fatal to the one who conquers it, Hagen's cup is mortal to the hero who drinks from it. A thousand stories and romances of this kind, both Germanic and Celtic, still haunt our sensibilities. Let us quote this stanza, in which Hreidmar, a hero of the *Edda* saga,[123] replies to the curse of Loki:

> You have given gifts,
> But you have not given gifts of love,
> You have not given with a kindly heart.
> You would already have been robbed of your life,
> If I had known earlier of the danger.

CELTIC LAW

Another family among Indo-European societies has certainly known such institutions. This is the family of Celtic peoples. Hubert and I have begun to provide the proof of this statement.[124]

CHINESE LAW

Finally, a great civilization, that of China, has retained from the most ancient times this very principle of law that is our concern. It acknowledges the indissoluble link that binds everything to its original owner. Even today an individual who has sold an item of his property,[125] even a movable good, preserves his whole life through, *vis-à-vis* the buyer, a kind of right 'to weep for his property'. Fr Hoang has transcribed models of these 'notes of complaint', which the seller hands to the buyer.[126] It is a kind of right of succession over the thing, mingled with a right of succession over the person, and which clings to the buyer a very long time after the thing has been definitively disposed of to another patrimony, and after all the terms of the 'irrevocable' contract have been carried out. Through the thing passed on, even if it is consumable, the alliance that has been contracted is no momentary phenomenon, and the contracting parties are deemed to be in a state of perpetual dependence towards one another.

In Annamite morality, to accept a present is dangerous. Westermarck,[127] who reports this fact, has perceived some of its importance.

4

CONCLUSION

MORAL CONCLUSIONS

It is possible to extend these observations to our own societies. A considerable part of our morality and our lives themselves are still permeated with this same atmosphere of the gift, where obligation and liberty intermingle. Fortunately, everything is still not wholly categorized in terms of buying and selling. Things still have sentimental as well as venal value, assuming values merely of this kind exist. We possess more than a tradesman morality. There still remain people and classes that keep to the morality of former times, and we almost all observe it, at least at certain times of the year or on certain occasions.

The unreciprocated gift still makes the person who has accepted it inferior, particularly when it has been accepted with no thought of returning it. We are still in the field of Germanic morality when we recall the curious essay by Emerson entitled 'Gifts'.[1] Charity is still wounding for him who has accepted it,[2] and the whole tendency of our morality is to strive to do away with the unconscious and injurious patronage of the rich almsgiver.

The invitation must be returned, just as 'courtesies' must. Surprisingly, here are to be seen traces of the old, traditional, moral basis, that of the ancient aristocratic potlatches. Here we also see come to the surface these fundamental motives for human activity: emulation between individuals of the same sex,[3] that 'basic imperialism' of human beings. On the one hand, it is the social basis, on the other the animal and psychological basis, that appears. In that separate existence that constitutes our social life, we ourselves cannot 'lag behind', as the expression still goes. We must give back more than we have received. The round of drinks is ever dearer and larger in size. Thus, in our childhood, one village family in Lorraine, which normally contented itself with living very frugally, ruined itself for the sake of its guests on saints days, and at weddings, first communions, or funerals. One must act the 'great lord' upon such occasions. It may even be said that one section of our people is constantly behaving like this, and spends with the utmost extravagance on guests and on feast days, and with New Year gifts.

The invitation must be given, and must be accepted. This is still the custom, even in our liberal society. Scarcely fifty years ago, and perhaps even more recently, in certain parts of France and Germany the entire village came to the wedding breakfast. If anyone stayed away it was a very bad omen, a foreboding, proof of envy, and a sign of bad luck. In France, in quite a number of places everybody still takes part in such ceremonies. In Provence, when a child is born, everybody still brings an egg and other symbolic presents.

Things sold still have a soul. They are still followed around by their former owner, and they follow him also. At Cornimont, in a valley of the Vosges, the following custom was common not so long ago and perhaps continues to linger on in certain families: so that animals that had been bought should forget their former master and were not tempted to return 'home', a cross was traced on the lintel of the stable door, the halter belonging to the seller was kept on the animals, and salt was fed to them. At Raon-aux-Bois the animals were given a slice of bread and butter that had been carried three times round the dairy and was held out to them with the right hand. It is true that this was only for the larger livestock that, since the stable was part of the house, were part of the family. But a number of other French customs denote that the thing sold must be detached from the seller, by, for example, striking the thing that is sold, or by whipping the sheep that is sold, etc.[4]

One might even say that a whole section of the law, that relating to industrialists and businessmen, is nowadays at odds with morality. The economic prejudices of the people, the producers, arise from their firm determination to follow the thing they have produced, and from the strong feeling they have that their handiwork is resold without their having had any share of the profit.

Nowadays the old principles react against the rigour, abstraction, and inhumanity of our legal codes. From this viewpoint it may be said that a whole section of our law that is just emerging, with certain customs, consists of turning back the clock. This reaction against the Roman and Saxon (sic) [Northern?] insensitivity of our system is perfectly healthy and well founded. A few new principles of law and custom may be interpreted in this way.

It took a considerable period of time to acknowledge proprietorship in artistic, literary, and scientific work, beyond the peremptory action of selling the manuscript, the first machine, or the original work of art. In fact, societies are not very interested in recognizing for the heirs of an author or an inventor – a benefactor of humanity – more than certain limited rights over the things created by the one that owns those rights. One likes to assert that they are the product of the collective mind as much as of individual mind. Everyone wishes them to fall into the public domain or join in the general circulation of wealth as quickly as possible. However, the scandal of the additional value acquired by paintings, sculptures, and objects d'art, during the lifetime of their creators or their immediate heirs, inspired the French law of September 1923 that gives to the artist and his inheritors a 'right of succession' over the series of additional gains made during the successive sales of their works.[5]

All our social insurance legislation, a piece of state socialism that has already been realized, is inspired by the following principle: the worker has given his life and his labour, on the one hand to the collectivity, and on the other hand, to his employers. Although the worker has to contribute to his insurance, those who have benefited from his services have not discharged their debt to him through the payment of wages. The state itself, representing the community, owes him, as do his employers, together with some assistance from himself, a certain security in life, against unemployment, sickness, old age, and death.

Even recent ingenious strategems, for example, the family assistance funds that our French industrialists have freely and energetically developed for the benefit of workers with family obligations, represent a spontaneous response to this need to forge links with individuals, to take into account the burdens they have to bear, and the varying degrees of material and moral interest that such burdens represent.[6] Similar bodies operate in Germany and Belgium with just as much success. In Britain, during this time of terrible, long drawn-out unemployment affecting millions of workers, an entire movement is emerging in favour of insurance against unemployment, which would be obligatory and organized through corporate bodies. The municipalities and the state are tired of shouldering this immense expense of payments to the unemployed, whose root cause lies in industry and market conditions alone. Thus distinguished economists and captains of industry such as Mr Pybus and Sir Lynden (sic) Macassey are urging that firms themselves, through their corporate associations, should organize these unemployment funds, and themselves make such sacrifices. In short, they would like the cost of security for the worker and his defence against being out of work to become part of the general expenses of each individual industry.

All such morality and legislation corresponds in our opinion, not to any upheaval in the law, but a return to it.[7] On the one hand, one is seeing the dawning, and even the realization, of professional morality and corporate law. The compensatory funds and mutual benefit societies that industrial groupings are setting up in order to finance corporate charitable works, from a purely moral viewpoint are entirely admirable, save on one score: they are run entirely by the employers. Moreover, it is groups that are acting: the state, the municipalities, institutions of public assistance, pension funds, savings banks, mutual benefit societies, employers, and wage-earners. They are all in association together, as, for example, under the social legislation existing in Germany and Alsace-Lorraine. Very soon they will be similarly associated in French social security schemes. Thus we are returning to a group morality.

On the other hand, the state and its subordinate grouping desire to look after the individual. Society is seeking to rediscover a cellular structure for itself. It is indeed wanting to look after the individual. Yet the mental state in which it does so is one in which are curiously intermingled a perception of the rights of the individual and other, purer sentiments: charity, social service, and solidarity. The themes of the gift, of the freedom and the obligation inherent in the gift, of generosity and self-interest that are linked in giving, are reappearing in French society, as a dominant motif too long forgotten.

But to note the fact is not enough. One must deduce practice from it, and a moral precept. It is not sufficient to say that the law is in the process of ridding itself of a few abstractions such as the distinction between real law and personal law; or that it is intent on adding other rights to the cold-hearted law of sale and payment for services. It must be said that this is a salutary revolution.

First of all, we return, as return we must, to habits of 'aristocratic extravagance'. As is happening in English-speaking countries and so many other contemporary societies, whether made up of savages or the highly civilized, the rich must come back to considering themselves – freely and also by obligation – as the financial guardians of their fellow citizens. Among ancient civilizations, from which ours has sprung, some had a (debtors') jubilee, others liturgies (of duty) such as choregies and trierarchies, and *syussitia* (meals in common), and the obligatory expenditure by the aedile and the consular dignitaries. We should return to laws of this kind. Then there must be more care for the individual, his life, his health, his education (which is, moreover, a profitable investment), his family, and their future. There must be more good faith, more sensitivity, more generosity in contracts dealing with the hiring of services, the letting of houses, the sale of vital foodstuffs. And it will indeed be necessary to find a way to limit the rewards of speculation and interest.

However, the individual must work. He should be forced to rely upon himself rather than upon others. On the other hand, he must defend his interests, both personally and as a member of a group. Over-generosity, or communism, would be as harmful to himself and to society as the egoism of our contemporaries and the individualism of our laws. In the *Mahabharata* a malevolent genie of the woods explains to a Brahmin who gave away too much, and too injudiciously: 'That is why you are thin and pale.' The life of the monk, and the life of a Shylock are both equally to be shunned. This new morality will surely consist of a good but moderate blend of reality and the ideal.

Thus we can and must return to archaic society and to elements in it. We shall find in this reasons for life and action that are still prevalent in certain societies and numerous social classes: the joy of public giving; the pleasure in generous expenditure on the arts, in hospitality, and in the private and public festival. Social security, the solicitude arising from reciprocity and co-operation, and that of the occupational grouping, of all those legal entities upon which English law bestows the name of 'Friendly Societies' – all are of greater value than the mere personal security that the lord afforded his tenant, better than the skimpy life that is given through the daily wages doled out by employers, and even better than capitalist saving – which is only based on a changing form of credit.

It is even possible to conceive what a society would be like in which such principles were the rule. In the liberal professions of our great nations to some extent a morality and an economy of this kind already flourish. For them honour, disinterestedness, corporate solidarity are not vain words, nor do they run counter to the necessities of work. Let us humanize in the same way other occupational groupings and improve them still further. This will represent great progress, as Durkheim has often advocated.

In so doing, we shall return, I think, to the enduring basis of law, to the very principle of normal social life. We must not desire the citizen to be either too good or too individualist nor too insensitive or too realist. He must have a keen sense of awareness of himself, but also of others, and of social reality (in moral matters is there even any other kind of reality?) He must act by taking into account his own interests, and those of society and its subgroups. This morality is eternal; it is common to the most advanced societies, to those of the immediate future, and to the lowest imaginable forms of society. We touch upon fundamentals. No longer are we talking in legal terms: we are speaking of men and groups of men, because it is they, it is society, it is the feelings of men, in their minds and in flesh and blood that at all times spring into action and that have acted everywhere.

Let us demonstrate this. The system that we propose to call the system of 'total services', from clan to clan – the system in which individuals and groups exchange everything with one another – constitutes the most ancient system of economy and law that we can find or of which we can conceive. It forms the base from which the morality of the exchange-through-gift has flowed. Now, that is exactly the kind of law, in due proportion, towards which we would like to see our own societies moving. To make these distant phases of law understood, here are two examples, borrowed from extremely different societies.

During a *corroboree* (a public drama dance) at Pine Mountain,[8] in mid-west Queensland, each individual in turn enters the con-secrated place, bearing in one hand his spear-slinger, and with his other hand behind his back. He throws his weapon from a circle at the other end of the dance area, at the same time calling out the place from where he comes, for example: 'Kunyan is my land'.[9] He stops for moment, and during this time his friends 'put a present', a spear, a boomerang, or some other weapon, into his other hand. 'A good warrior can thus receive more than his hand can hold, particularly if he has daughters to marry off.'[10]

In the Winnebago tribe (the Sioux tribe), the chiefs of the clans very typically give speeches to their fellow chiefs[11] from other tribes that are models of that etiquette[12] widespread in all the Indian civilizations of North America. Each clan cooks food and prepares tobacco for the representatives of the other tribes, during the clan's festival. Here, for example, are excerpts from the speech made by the chief of the Snake clan:[13]

> I greet you. It is good. How could I do otherwise? I am a poor, worthless man and you have remembered me. It is good . . . You have thought of the spirits and you have come to sit down with me . . . Soon your dishes will be filled. So I greet you once again, you humans that take the place of the spirits. Etc.

And when each chief has eaten, and has put offerings of tobacco into the fire, the final form of words sets out the moral effect of the festival, and all the services that have been rendered:

> I thank you for having come to sit down in this seat, I am grateful to you. You have encouraged me . . . The blessings of your grandfathers who have enjoyed revelations (and who are incarnate in you) are equal to those of the spirits. It is good that you have taken part in my festival. As our ancestors have said, this must be: 'Your life is weak and you can only be strengthened by the counsel of the braves.' You have counselled me . . . This is life for me.

Thus, from one extreme of human evolution to the other, there are no two kinds of wisdom. Therefore let us adopt as the principle of our life what has always been a principle of action and will always be so: to emerge from self, to give, freely and obligatorily. We run no risk of disappointment. A fine Maori proverb runs:

> Ko Maru kai atu
> Ko maru kai mai
> ka ngohe ngohe.

'Give as much as you take, all shall be very well.'[14]

II
CONCLUSIONS FOR ECONOMIC SOCIOLOGY AND POLITICAL ECONOMY

These facts not only throw light upon our morality and help to direct our ideals. In their light, we can analyse better the most general economic facts, and even this analysis helps us dimly to perceive better organizational procedures applicable in our societies.

Several times we have seen how far this whole economy of the exchange-through-gift lay outside the bounds of the so-called natural economy, that of utilitarianism. All these very considerable phenomena of the economic life of all peoples – let us say, to fix things firmly in our minds, that they represent fittingly the great Neolithic civilization – and all these important vestiges of those traditions in societies close to our own, or of our own customs, fall outside the schemes normally put forward by those rare economists who have wished to compare the various types of known economies.[15] We therefore add our own repeated observations to those of Malinowski, who has devoted an entire study to 'exploding' current doctrines concerning 'primitive' economy.[16]

From this there follows a very solid chain of facts: the notion of value functions in these societies. Very large surpluses, speaking in absolute terms, are amassed. They are often expended to no avail, with comparatively enormous luxury,[17] which is in no way commercial. These are the signs of wealth, and kinds of money[18] are exchanged. Yet the whole of this very rich economy is still filled with religious elements. Money still possesses its magical power and is still linked to the clan or to the individual.[19] The various economic activities, for example the market, are suffused with rituals and myths. They retain a ceremonial character that is obligatory and effective.[20] They are full of rituals and rights. In this light we can already reply to the question that Durkheim posed concerning the religious origin of the notion of economic value.[21] The facts also answer a host of questions concerning the forms and reasons behind what we so ineptly term exchange, the 'barter', the *permutation*[22] of useful things, that, in the wake of the prudent Romans, who were themselves following Aristotle,[23] an *a priori* economic history places at the origin of the division of labour. It is indeed something other than utility that circulates in societies of all kinds, most of which are already fairly enlightened. The clans, the generations, and the sexes generally – because of the many different relationships to which the contracts give rise – are in a perpetual state of economic ferment and this state of excitement is very far from being materialistic. It is far less prosaic than our buying and selling, our renting of services, or the games we play on the Stock Exchange.

However, we can go even farther than we have gone up to now. One can dissolve, jumble up together, colour and define differently the principal notions that we have used. The terms that we have used – present and gift – are not themselves entirely exact. We shall, however, find no others. These concepts of law and economics that it pleases us to contrast: liberty and obligation; liberality, generosity, and luxury, as against savings, interest, and utility – it would be good to put them into the melting pot once more. We can only give the merest indications on this subject. Let us choose,[24] for example, the Trobriand Islands. There they still have a complex notion that inspires all the economic acts we have described. Yet this notion is neither that of the free, purely gratuitous rendering of total services, nor that of production and exchange purely interested in what is useful. It is a sort of hybrid that flourished.

Malinowski has made a serious attempt at classifying,[25] from the point of view of motives of self-interest and disinterestedness, all the transactions that he noted among the Trobriand Islanders. He gradates them between the pure gift and pure barter after bargaining has taken place.[26] This classification is in reality inapplicable. Thus, according to Malinowski, the type of pure gift would be the gift between man and wife.[27] But, in our view, precisely one of the most important facts reported by Malinowski and one that throws a brilliant light upon all sexual relationships throughout humanity, consists of comparing the *mapula*,[28] the 'constant' payment made by the man to his wife, as a kind of salary for sexual services rendered.[29] Likewise the presents made to the chief are a tribute paid; the distributions of food (*sagali*) are rewards for work or rituals performed, for example, in the case of funeral vigils.[30] All in all, just as these gifts are not freely given, they are also not really disinterested. They already represent for the most part total counter-services, not only made with a view to paying for services or things, but also to maintaining a profitable alliance,[31] one that cannot be rejected. Such, for example, is the alliance between tribes of fishermen[32] and tribes of farmers or pottery-makers. Now, this is a general fact. We have met it, for example, in the Maori and Tsimshian areas, etc.[33] We can therefore see where this force resides. It is one that is both mystical and practical, one that ties clans together and at the same time divides them, that divides their labour, and at the same time constrains them to carry out exchange. Even in these societies, the individual and the group, or rather the subgroup, have always felt they had a sovereign right to refuse a contract. It is this that gives the stamp of generosity to this circulation of goods. On the other hand they normally had neither the right to, nor any interest in refusing. It is this that makes these distant societies nevertheless related to our own.

The use of money might suggest other reflections. The *vaygu'a* of the Trobriands, bracelets and necklaces, just as the copper objects of the American Northwest or the *wampun* of the Iroquois, are both riches, signs of wealth,[34] and means of exchange and of payment, but also things that must be given, or even destroyed. However, these are still pledges linked to the persons that use them, and these pledges bind them. Since, on the other hand, they already serve as indicators of money, one has an interest in giving them away so as to be able to possess yet other objects, by transforming them into goods or services that, in their turn, can be transformed again into money. One might really say that the Trobriand or Tsimshian, although far removed from him, proceeds like the capitalist who knows how to dispose of his ready cash at the right time, in order to reconstitute at a later date this mobile form of capital. Self-interest and disinterestedness like-wise explain this form of the circulation of wealth and that of the archaic circulation of the signs of wealth that ensue.

Even pure destruction of wealth does not signify that com-plete detachment that one might believe to be found in it. Even these acts of greatness are not without egoism. The purely sump-tuary form of consumption (which is almost always exaggerated and often purely destructive), in which considerable amounts of goods that have taken a long time to amass are suddenly given away or even destroyed, particularly in the case of the potlatch,[35] give such institutions the appearance of represent-ing purely lavish expenditure and childish prodigality. In effect, and in reality, not only are useful things given away and rich foods consumed to excess, but one even destroys for the pleasure of destroying. For example, the Tsimshian, Tlingit, and Haïda chiefs throw these copper objects and money into the water. The Kwakiutl chiefs smash them, as do those of the tribes allied to them. But the reason for these gifts and frenetic acts of wealth consumption is in no way disinterested, particu-larly in societies that practise the potlatch. Between chiefs and their vassals, between vassals and their tenants, through such gifts a hierarchy is established. To give is to show one's superiority, to be more, to be higher in rank, *magister*. To accept without giving in return, or without giving more back, is to become client and servant, to become small, to fall lower (*minister*).

The magic ritual of the *kula* called the *mwasila*[36] is full of formu-las and symbols that demonstrate that the potential contracting party seeks above all this advantage of social superiority – one might almost say brute superiority. Thus, having cast a spell over the betel nut that they are going to use with their partners, after having cast a spell over the chief, his comrades, their pigs, the necklaces, then the head and its orifices, plus everything that is brought there, the *pari*, the opening gifts, etc. . . . after having cast a spell over all these, the magician sings, not without exaggeration:[37]

> I topple the mountain, the mountain moves, the mountain crumbles away, etc. . . . My charm goes to the summit of the Dobu mountain . . . My boat is going to sink . . . etc. My fame is like that of the lightning. My tread is like that of the flying witch doctors, the Tudududu.

To be first, the most handsome, the luckiest, the strongest, and wealthiest – this is what is sought after, and how it is obtained. Later, the chief gives proof of his *mana* by redistributing what he has just received to his vassals and relations. He sustains his rank among the chiefs by giving back bracelets for necklaces, hospitality for visits, etc. In this case riches are from every viewpoint as much a means of retaining prestige as something useful. Yet are we sure that it is any different in our own society, and that even with us riches are not above all a means of lording it over our fellow men?

Let us now put to the test the other notion that we have just opposed to that of the gift and disinterestedness: the notion of interest, of the individual search after what is useful. This does not present itself either as it functions in our own minds. If some equivalent reason animates the Trobriand or American Indian chiefs, the Andaman clans, etc., or once motivated generous Hindus, and Germanic or Celtic nobles, as regards their gifts and expenditure, it is not the cold reasoning of the merchant, the banker, and the capitalist. In those civilizations they are concerned with their own interest, but in a different way from our own age. They hoard, but in order to spend, to place under an obligation, to have their own 'liege men'. On the other hand, they carry on exchange, but it is above all in luxury articles, ornaments or clothes, or things that are consumed immediately, as at feasts. They repay with interest, but this is in order to humiliate the person initially making the gift or exchange, and not only to recompense him for loss caused to him by 'deferred consumption'. There is self-interest, but this self-interest is only analogous to what allegedly sways us.

A relatively amorphous and disinterested economic system exists within subgroups, that regulates the life of the Australian clans or those of North America (the East and the Prairies). On the other hand there exists also the individualistic and purely self-interested economy that our own societies have experienced at least in part, as soon as it was discovered by the Semitic and Greek peoples. Between these two types there is an entire and immensely gradated series of institutions and economic events, and this series is not governed by the economic rationalism whose theory we are so willing to propound.

The very word 'interest' is itself recent, originally an accounting technique: the Latin word *interest* was written on account books against the sums of interest that had to be collected. In ancient systems of morality of the most epicurean kind it is the good and pleasurable that is sought after, and not material utility. The victory of rationalism and mercantilism was needed before the notions of profit and the individual, raised to the level of principles, were introduced. One can almost date − since Mandeville's *The Fable of the Bees* − the triumph of the notion of individual interest. Only with great difficulty and the use of periphrasis can these two words be translated into Latin, Greek, or Arabic. Even those who wrote classical Sanskrit, who used the word *artha*, fairly close to our own idea of interest, had a different idea of it from our own, as they did for other categories of action. The sacred books of classical India already divide human activities up as follows: law (*dharma*), interest (*artha*), desire (*kama*). But above all it is a matter of *political* self-interest − that of the king and the Brahmins, of the ministers, in the kingdom and in each caste. The considerable literature of the *Niticastra* is not concerned with economics.

It is our western societies who have recently made man an 'economic animal'. But we are not yet all creatures of this genus. Among the masses and the elites in our society purely irrational expenditure is commonly practised. It is still characteristic of a few of the fossilized remnants of our aristocracy. *Homo oeconomicus* is not behind us, but lies ahead, as does the man of morality and duty, the man of science and reason. For a very long time man was something different, and he has not been a machine for very long, made complicated by a calculating machine.

Moreover, happily we are still somewhat removed from this constant, icy, utilitarian calculation. We need to carry out an analysis in depth, with statistics, as Halbwachs has done for the working classes, of our own consumption and expenditure, we of the western middle class. How many needs do we satisfy? And how many inclinations do we not satisfy whose ultimate purpose is not one of utility? How much of his income does or can the rich man allocate to his personal utilitarian needs? His expenditure on luxury, on art, on outrageous things, on servants – do not these make him resemble the nobles of former times or the barbarian chiefs whose customs we have described?

Is it good that this should be so? That is a different question. It is perhaps good that there are other means of spending or exchanging than pure expenditure. In our view, however, it is not in the calculation of individual needs that the method for an optimum economy is to be found. I believe that we must remain something other than pure financial experts, even in so far as we wish to increase our own wealth, whilst becoming better accountants and managers. The brutish pursuit of individual ends is harmful to the ends and the peace of all, to the rhythm of their work and joys – and rebounds on the individual himself.

As we have just seen, already important sections of society, associations of our capitalist firms themselves, are seeking as bodies to group their employees together. Moreover, all syndical-ist groupings, whether of employers or wage-earners, claim they are defending and representing the general interest as fervently as the individual interest of their members or even their corporations. These fine speeches, it is true, are adorned with many metaphors. However, we must state that not only morality and philosophy, but even public opinion and political economy itself, are beginning to elevate themselves to this 'social' level. We sense that we cannot make men work well unless they are sure of being fairly paid throughout their life for work they have fairly carried out, both for others and for themselves. The producer who carries on exchange feels once more – he has always felt it, but this time he does so acutely – that he is exchanging more than a product of hours of working time, but that he is giving something of himself – his time, his life. Thus he wishes to be rewarded, even if only moderately, for this gift. To refuse him this reward is to make him become idle or less productive.

Perhaps we may point out a conclusion that is both sociological and practical. The famous Sourate LXIV, 'mutual disappointment' (the Last Judgement) given to Mahomet at Mecca, says of God:

> 15. Your wealth and your children are your temptation, whilst God holds in reserve a magnificent reward.
> 16. Fear God with all your might; listen and obey, give alms (*sadaqa*) in your own interest. He who is on his guard against his avarice will be happy.
> 17. If you make a generous loan to God, he will pay you back double; he will forgive you because he is grateful and long-suffering.
> 18. He knows things visible and invisible, he is the one powerful and wise.

Substitute for the name of Allah that of society and the occupational grouping, or put together all three names, if you are religious. Replace the concept of alms by that of co-operation, of a task done or service rendered for others. You will then have a fairly good idea of the kind of economy that is at present laboriously in gestation. We see it already functioning in certain economic groupings, and in the hearts of the masses, who possess, very often better than their leaders, a sense of their own interests, and of the common interest.

Perhaps by studying these obscure aspects of social life we shall succeed in throwing a little light upon the path that our nations must follow, both in their morality and in their economy.

III
CONCLUSION REGARDING GENERAL SOCIOLOGY AND MORALITY

May we be allowed a further remark about the method we have followed? We have no wish to put forward this study as a model to be followed. It only sets out bare indications. It is not sufficiently complete and the analysis might be pushed still farther.[38] We are really posing questions to historians and ethnographers, and putting forward subjects for enquiry rather than resolving a problem and giving a definitive answer. For the time being it is enough for us to be persuaded that in this direction numerous facts will be discovered.

Yet, if this is so, it is because in this way of treating a problem there lies a heuristic principle we should like to bring out. The facts that we have studied are all, if we may be allowed the expression, *total* social facts, or, if one wishes – although we do not like the word – general ones. That is to say, in certain cases they involve the totality of society and its institutions (potlatch, clans confronting one another, tribes visiting one another, etc.), and in other cases only a very large number of institutions, particularly when these exchanges and contracts rather concern the individual.

All these phenomena are at the same time juridical, economic, religious, and even aesthetic and morphological, etc. They are juridical because they concern private and public law, and a morality that is organized and diffused throughout society; they are strictly obligatory or merely an occasion for praise or blame; they are political and domestic at the same time, relating to social classes as well as clans and families. They are religious in the strict sense, concerning magic, animism, and a diffused religious mentality. They are economic. The idea of value, utility, self-interest, luxury, wealth, the acquisition and accumulation of goods – all these on the one hand – and on the other, that of consumption, even that of deliberate spending for its own sake, purely sumptuary: all these phenomena are present everywhere, although we understand them differently today. Moreover, these institutions have an important aesthetic aspect that we have deliberately omitted from this study. Yet the dances that are carried out in turn, the songs and processions of every kind, the dramatic performances that are given from camp to camp, and by one associate to another; the objects of every sort that are made, used, ornamented, polished, collected, and lovingly passed on, all that is joyfully received and successfully presented, the banquets themselves in which everyone participates; everything, food, objects, and services, even 'respect', as the Tlingit say, is a cause of aesthetic emotion, and not only of emotions of a moral order or relating to self-interest.[39] This is true not only for Melanesia, but even more especially so for the system of potlatch in the American Northwest, and still more so for the festival-cum-market of the Indo-European world.[40] Finally, the phenomena are clearly structural. They all occur during assemblies, fairs, and markets, or at least at festivals that take their place. All such festivals presuppose congregations whose duration can exceed one season of social coming together, as do the winter potlatches of the Kwakiutl, or weeks, as do the seafaring expeditions of the Melanesians. Moreover, there must be roads or at least trails, and seas and lakes across which one may peaceably transport oneself. There must be tribal, intertribal, or international alliances, those of the *commercium* and the *connubium*.[41]

Thus these are more than themes, more than the bare bones of institutions, more than complex institutions, even more than systems of institutions divided, for example, into religion, law, economy, etc. They are whole 'entities', entire social systems, the functioning of which we have attempted to describe. We have looked at societies in their dynamic or physiological state. We have not studied them as if they were motionless, in a static state, or as if they were corpses. Even less have we decomposed and dissected them, producing rules of law, myths, values, and prices. It is by considering the whole entity that we could perceive what is essential, the way everything moves, the living aspect, the fleeting moment when society, or men, become sentimentally aware of themselves and of their situation in relation to others. In this concrete observation of social life lies the means of discovering new facts, which we are only beginning dimly to perceive. In our opinion, nothing is more urgent or more fruitful than this study of total social facts.

It has a double advantage. Firstly, there is the advantage of generality. Those facts that relate to the general functioning of society are likely to be more universal than the various institutions, or the various themes that relate to these institutions, which are always more or less accidentally tinged with local colour. But above all such a study has the advantage of reality. Thus one succeeds in seeing the social 'things' themselves, in concrete form and as they are. In societies one grasps more than ideas or rules, one takes in men, groups, and their different forms of behaviour. One sees them moving, as one does masses and systems in mechanics, or as in the sea we notice the octopuses and the anemones. We perceive numbers of men, forces in motion, who are in movement in their environment and in their feelings.

The historians feel and rightly object to the fact that the sociologists are too ready with abstractions and unduly separate the various elements of societies from one another. We must do as they do: observe what is given. Now, the given is Rome or Athens, the average Frenchmen, the Melanesian from this island or another, and not prayer or law by itself. After having of necessity divided things up too much, and abstracted from them, the sociologists must strive to reconstitute the whole. By so doing they will discover rewarding facts. They will also find a way to satisfy the psychologists. The latter are strongly aware of their privileged position; the psychopathologists, in particular, are certain that they can study the concrete. All these study or should observe, the behaviour of total beings, not divided according to their faculties. We must imitate them. The study of the concrete, which is the study of completeness, is possible, and more captivating, more explanatory still in sociology. For our part, we observe the complete and complex reactions of numerically defined masses of men, complete, complex beings. We, too, observe what constitutes their organism or their psyche. At the same time we describe the behaviour of this mass and its corresponding psychoses: sentiments, ideas, and the volitions of the crowd, or of organized societies and their subgroups. We, too, view entities, and the reactions of those entities, whose ideas and feelings are normally interpretations, and more rarely do we see the reasons for them. The principle and the end of sociology is to perceive the whole group and its behaviour in its entirety.

We have not had time – it would have meant extending unduly a limited subject – to try to perceive at this time the structural basis for all the facts we have indicated. However, it is perhaps useful to indicate, at least as an example, the method we would like to follow, and along what lines we would carry out that research.

All the societies, we have described above, except for our European societies, are segmented. Even Indo-European societies – Roman society before the Twelve Tables, Germanic societies even very late on, up to the writing down of the *Edda* saga, and Irish society up to the creation of its main literature – were still based on the clan and, at the very least the large families, which formed internally a more or less undivided block, being more or less externally isolated from one another. All these societies are or were far from our own state of unification, or the unity that a defective history ascribes to them. Moreover, within these groups, individuals, even those with strong characteristics, were less sad, less serious, less miserly, and less personal than we are. Externally at least, they were or are more generous, more liable to give than we are. The law of friendship and contracts, with the gods, came to ensure 'peace' within 'markets' and towns. This occurred when groups paid visits to one another at tribal festivals and at ceremonies where clans confronted one another and families allied themselves or began 'initiations' with one another. It happened even in more advanced societies when the 'law of hospitality' had been developed. Over a considerable period of time and in a considerable number of societies, men approached one another in a curious frame of mind, one of fear and exaggerated hostility, and of generosity that was likewise exaggerated, but such traits only appear insane to our eyes. In all the societies that have immediately preceded our own, and still exist around us, and even in numerous customs extant in our popular morality, there is no middle way: one trusts completely, or one mistrusts completely; one lays down one's arms and gives up magic, or one gives everything, from fleeting acts of hospitality to one's daughter and one's goods. It is in such a state of mind that men have abandoned their reserve and have been able to commit themselves to giving and giving in return.

This was because they had no choice. Two groups of men who meet can only either draw apart, and, if they show mistrust towards one another or issue a challenge, fight – or they can negotiate. Until legal systems and economies evolved not far removed from our own, it is always with strangers that one 'deals', even if allied to them. In the Trobriand Islands the people of Kiriwina told Malinowski:[42] 'The men from Dobu are not good like us; they are cruel, they are cannibals. When we come to Dobu, we are afraid of them. They might kill us. But then I spit out ginger root, and their attitude changes. They lay down their spears and receive us well.' Nothing better interprets this unstable state between festival and war.

One of the best ethnographers, Thurnwald, writing about another Melanesian tribe, giving genealogical statistics,[43] describes for us a particular event that also clearly demonstrates how these people, as a group, suddenly pass from festival to battle. Buleau, a chief, had invited another chief, Bobal, and his people to a banquet, probably the first in a long series. They began to rehearse the dances the whole night through. In the morning they were all in a state of nerves from their sleepless night, the dances, and the songs. As a result of a simple remark made by Buleau, one of Bobal's men killed him. And the rank and file massacred, pillaged, and carried off the women of the village. 'Buleau and Bobal were rather friendly, and merely rivals', Thurnwald was told. We have all observed such facts, even around us.

It is by opposing reason to feeling, by pitting the will to peace against sudden outbursts of insanity of this kind that peoples succeed in substituting alliance, gifts, and trade for war, isolation and stagnation.

This is therefore what one may have found at the conclusion of this research. Societies have progressed in so far as they themselves, their subgroups, and lastly, the individuals in them, have succeeded in stabilizing relationships, giving, receiving, and finally, giving in return. To trade, the first condition was to be able to lay aside the spear. From then onwards they succeeded in exchanging goods and persons, no longer only between clans, but between tribes and nations, and, above all, between individuals. Only then did people learn how to create mutual interests, giving mutual satisfaction, and, in the end, to defend them without having to resort to arms. Thus the clan, the tribe, and peoples have learnt how to oppose and to give to one another without sacrificing themselves to one another. This is what tomorrow, in our so-called civilized world, classes and nations and individuals also, must learn. This is one of the enduring secrets of their wisdom and solidarity.

There is no other morality, nor any other form of economy, nor any other social practices save these. The Bretons, and the *Chronicles of Arthur*[44] tell how King Arthur, with the help of a Cornish carpenter, invented that wonder of his court, the miraculous Round Table, seated round which, the knights no longer fought. Formerly, 'out of sordid envy', in stupid struggles, duels and murders stained with blood the finest banquets. The carpenter said to Arthur: 'I will make you a very beautiful table, around which sixteen hundred and more can sit, and move around, and from which no-one will be excluded . . . No knight will be able engage in fighting, for there the highest placed will be on the same level as the lowliest.' There was no longer a 'high table', and consequently no more quarrelling. Everywhere that Arthur took his table his noble company remained happy and unconquerable. In this way nations today can make themselves strong and rich, happy and good. Peoples, social classes, families, and individuals will be able to grow rich, and will only be happy when they have learnt to sit down, like the knights, around the common store of wealth. It is useless to seek goodness and happiness in distant places. It is there already, in peace that has been imposed, in well-organized work, alternately in common and separately, in wealth amassed and then redistributed, in the mutual respect and reciprocating generosity that is taught by education.

In certain cases, one can study the whole of human behaviour, and social life in its entirety. One can also see how this concrete study can lead not only to a science of customs, to a partial social science but even to moral conclusions, or rather, to adopt once more the old word, 'civility', or 'civics', as it is called nowadays. Studies of this kind indeed allow us to perceive, measure, and weigh up the various aesthetic, moral, religious, and economic motivations, the diverse material and demographic factors, the sum total of which are the basis of society and constitute our common life, the conscious direction of which is the supreme art, Politics, in the Socratic sense of the word.

Marcel Mauss

Notes

INTRODUCTION

1 Cassel, in his *Theory of Social Economy*, vol. 2, p. 345, put us on the track of this text. Scandinavian scholars are familiar with this trait of the ancient history of their peoples.

2 Maurice Cahen kindly agreed to make this translation for us.

3 The meaning of the stanza is obscure, particularly because the adjective is missing in line 4, but the drift is clear when, as is normally done, one supplies a word that means 'liberal', or 'extravagant'. Line 3 is also difficult. Cassel translates it: 'who does not take what is offered him'. By contrast, Cahen's translation is literal. He has written to us: 'The expression is ambiguous. Some understand it to mean: "that to receive would not be pleasing to him". Others interpret it as: "that to receive a present would not carry an obligation to reciprocate". Naturally I incline to the second explanation.' In spite of our lack of expertise in Old Norse, we venture to put forward another interpretation. Clearly the expression corresponds to an old cento that must have run something like, 'to receive is received'. If this is conceded, the line would allude to the state of mind of the visitor and his host. Everyone is supposed to offer hospitality, or his presents, as if they were never to be reciprocated. However, everyone nevertheless

accepts the visitor's presents or the total gifts and services provided by the host in return, because they represent property and are also a means of strengthening the contract, of which they form an integral part.

For us, it even appears that in these stanzas can be discerned a much older version. The structure of each one is the same, curious but clear. In each a juridical cento forms the centrepiece: 'that to receive would not be received' (stanza 39), 'those that exchange presents are friends' (41), 'to give present for present' (42), 'your soul must blend in with his and you must exchange presents' (44), 'the miser is always afraid of presents' (48), 'a present given always expects a present in return' (145), etc. It is a veritable collection of sayings. This proverb, or rule, is embedded in a commentary that embroiders on it. So we have here not only a very ancient form of law, but even a very ancient form of literature.

4   I have not been able to consult Burckhard, *Zum Begriff der Schenkung*, p. 53 ff. However, for Anglo-Saxon law, the fact that we propose to shed light on was very clearly appreciated by Pollock and Maitland, *History of English Law*, vol. 2, p. 892: 'The wide word "gift", which will cover sale, exchange, gage and lease.' Cf. ibid, p. 12, ibid, pp. 212–14: 'There is no free gift that carries the force of law.' See also the dissertation by Neubecker (1909) concerning Germanic dowry, *Die Mitgift*, p. 65 ff.

5   The notes are only indispensable to specialists.

6   G. Davy (1922) 'Foi jurée', *Travaux de l'Année Sociologique*; for bibliographical information see: M. Mauss (1921) 'Une forme archaïque de contrat chez les Thraces', *Revue des etudes grecques*; R. Lenoir (1924) 'L'Institution du Potlach', *Revue Philosophique*.

7   M.F. Sondo (1909) *Der Güterverkehr in der Urgesellschaft* (Institut Solvay), has provided a good discussion of these facts, and a clue (p. 156) that he was beginning to follow the path that we ourselves will follow.

8   Grierson (1903) *Silent Trade*, has already put forward the necessary arguments to rid ourselves of this prejudice. So has von Moszkowski (1911) *Vom Wirtschaftsleben der primitiven Völker*, although he considers theft as a primitive phenomenon and finally confuses the right to take something with theft. A fine exposition of the facts regarding the Maoris is to be found in W. von Brun [sic] (1912) *Wirtschaftsorganisation der Maori* (Lamprecht's contribution, p. 18), Leipzig, in which a chapter is devoted to exchange. The most recent summary of work on

the economy of peoples considered to be primitive is: Koppers (1915–16) 'Ethnologische Wirtschaftsord-nung', *Anthropos*, pp. 611–51, 971–1079. This is particularly good as an exposition of the doctrines held, but otherwise somewhat dialectic.

9   Since our most recent publications we have discovered in Australia the beginnings of regulated 'total services' occurring between tribes, particularly upon the occasion of a death. Furthermore, it is no longer solely between clans and phratries. Among the Kakado of the Northern Territory, a third funeral ceremony follows upon the second burial. During this ceremony the men proceed to a kind of judicial inquiry in order to determine, at least nominally, the one who had by sorcery perpetrated the death. But contrary to what follows in most Australian tribes, no vendetta is embarked upon. The men confine themselves to collecting up their spears and to working out what they will ask for in exchange. The next day these spears are taken away by another tribe – the Umoriu, for example – in whose camp the purpose for which they have been sent is perfectly understood. There the spears are laid out in heaps according to their owners, and, in accordance with a tariff known beforehand, the objects that are wanted are laid down opposite these heaps. Then everything is taken back to the Kakadu (Baldwin Spencer (1914) *Tribes of the Northern Territory*, p. 247). Sir Baldwin mentions that these objects can be exchanged once more against the spears, a fact that we do not understand very clearly. On the contrary, he finds it difficult to grasp the connection between these funeral rites and the exchanges, and he adds that the natives have no idea of the reason. Nevertheless the custom is perfectly comprehensible: it is to some extent a juridical settlement arrived at according to rules, which replaces the vendetta and serves as the origin for an intertribal market. This exchange of things is at the same time the exchange of pledges of peace and solidarity in mourning, as takes place normally in Australia among clans and families associated with one another and related by marriage. The sole difference is that this time the custom has become one between tribes.

10 But such a late-classical poet as Pindar says: βεαβιφ υανψρφ πραπιβψβ οιξοσεβ οιξαδε, *Olympic*, 8, 4. The entire passage is still imbued with the state of law that we shall describe. The themes of the present, of wealth, marriage, honour, favour, alliance, the shared meal, and the consecrated drink, even that of the jealousy that marriage arouses – all are represented in the passage in expressive terms that deserve comment.

11   See especially the remarkable rules for a ball game among the Omaha: Alice Fletcher and La Flesche (1905–6) 'Omaha Tribe', *Annual Report of the Bureau of American Anthropology* 27: 197, 366.

12   Krause, *Tlinkit Indianer*, p. 234 ff., clearly perceived this characteristic of the festivities and rituals that he describes, without calling them a potlatch. Boursin, in Porter (1900) 'Report on the Population . . . of Alaska', in *Eleventh Census*, pp. 54–66, and Porter, ibid, p. 33 did in fact perceive this feature of mutual glorification in the potlatch, which this time is named as such. But it is Swanton (1905) who has most clearly stressed it in: 'Social Conditions of the Tlingit Indians', *Annual Report of the Bureau of American Ethnography*, 26:345, etc. See our own observations in *Année Sociologique* 11: 207 and G. Davey (1922) 'Foi jurée', p. 172.

13   On the meaning of the word potlatch see Barbeau (1911) *Bulletin de la Société de Géographie de Québec*, and Davey, p. 162. However, it does not seem to us that the meaning propounded is the original one. In fact Boas gives the word potlatch – in Kwakiutl, it is true, and not in Chinook – the meanings of 'feeder', and literally, 'place of being satiated': *Kwakiutl Texts*, second series, Jesup [sic] Expedition, vol. 10, p. 43, n. 2; see also ibid, vol. 3, pp. 255, 517 under the heading *PoL*. But the two meanings of potlatch, i.e. 'gift' and 'food' are not mutually exclusive, since at least in theory the essential form of the 'total service' relates to nourishment. On this meaning see Ch. 2, n. 209 (p. 122).

14   The juridical aspect of the potlatch has been studied by Adam in his articles in the *Zeitschrift für vergleichender Rechtswissenschaft*, from 1911 onwards, and the *Festschrift*, Seler (1920), as well as Davy (1922) in his 'Foi jurée'. The religious and economic aspects are no less essential and should be treated equally thoroughly. The religious nature of the persons involved and of the things exchanged or destroyed are indeed not irrelevant to the very nature of contracts, no less the values that are ascribed to them.

15   The Haïda speak of 'killing' wealth.

16   See Hunt's documents in Boas, 'Ethnology of the Kwakiutl', *Annual Report of the Bureau of American Ethnography*, 35, 2: 1340, in which is to be found an interesting description of the way in which the clan brings its contributions for the potlatch to the chief, and also some very interesting palaver. In particular, the chief says, 'For it will not be in my name. It will be in your name and you will become famous among the tribes when it is said that you are giving your property for a potlatch.' (p. 1372, line 34 ff.)

17 In fact the domain of the potlatch extends beyond the limits of the Northwest tribes. In particular, one must consider whether the 'asking festival' of the Alaskan Eskimos is anything more than a borrow from the neighbouring Indian tribes. See Ch. 1, n. 45, p. 93.

18 See our remarks in *Année Sociologique* 11: 101; 12: 372–4, and *Anthropologie* (1920) (report of the sessions of the Institut Français d'Anthropologie). Lenoir (1924) has pointed out two fairly clear cases of potlatch in South America ('Expéditions maritimes en Mélanesie', *Anthropologie*, September.)

19 The word is used by M. Thurnwald (1912) *Forschungen auf den Salomo-Inseln* vol. 3, p. 8.

20 Mauss, M. (1921) *Revue des études grecques*, vol. 34.

1 THE EXCHANGE OF GIFTS AND OBLIGATION TO RECIPROCATE (POLYNESIA)

1 G. Davy (1922) 'Foi jurée', p. 140, has studied these exchanges in connection with marriage, and its relationship to contract. As we shall see, they have a different dimension.

2 Turner, *Nineteen Years in Polynesia*, p. 178; *Samoa*, p. 82 ff.; Stair, *Old Samoa*, p. 175.

3 Krämer, *Samoa-Inseln*, vol. 2, pp. 52–63.

4 Stair, *Old Samoa*, p. 180; Turner, *Nineteen Years in Polynesia*, p. 225; *Samoa*, p. 142.

5 Turner, *Nineteen Years in Polynesia*, p. 184; *Samoa*, p. 91.

6 Krämer *Samoa-Inseln*, vol. 2, p. 105; Turner, *Samoa*, p. 142.

7 Krämer, *Samoa-Inseln*, vol. 2, pp. 96, 363. The commercial expedition, the *malaga* (cf. *walaga* in New Guinea) corresponds in fact very closely to the potlatch, which itself is characteristic of the expeditions carried out in the neighbouring Melanesian archipelago. Krämer uses the word *Gegenschenk* ['reciprocating present'] for the exchange of the *oloa* against the *tonga*, which we shall discuss. Moreover, although we must not fall into the exaggerations of British ethnographers of the Rivers and Elliot Smith school, nor into those of American ethnographers who, following Boas, see the whole of the American system of potlatch as a series of borrowings, we should, however, lay much weight on the fact that institutions, so to speak, travel around. This is especially true in this case, where a considerable amount of trade, from island to island and port to port, and over very great distances, from very early times must have served not only the

passage of goods, but also the ways in which they were exchanged. Malinowski, in studies that we shall cite later, had a judicious appreciation of this fact. Cf. a study devoted to some of these institutions (Northwest Melanesia), in R. Lenoir (1924) 'Expéditions maritimes en Mélanesie', *Anthropologie*, September.

8   In any case rivalry between Maori clans is mentioned fairly often, particularly in connection with festivities. Cf. S.P. Smith, *Journal of the Polynesian Society* (henceforth, *JPS*), vol. 15, p. 87. (See also pp. 1, 59, n. 4).

9   The reason why, in this case, we do not assert that potlatch proper exists, is because the element of usury in the reciprocal service rendered is lacking. However, as we shall see in considering Maori law, the fact that nothing is given in return entails the loss of *mana*, of 'face', as the Chinese say. In Samoa also, in order not to incur the same disadvantage, 'give and give in return' must be observed.

10 Turner, *Nineteen Years in Polynesia*, p. 178; *Samoa*, p. 52. This theme of ruin and honour is a basic one in the potlatch of the American Northwest. Cf. examples in Porter, 'Report . . . ', *Eleventh Census*, p. 334.

11 Turner, *Nineteen Years in Polynesia*, p. 178; *Samoa*, p. 83, calls the young man 'adopted'. He is wrong. The custom is exactly that of 'fosterage', of education being given outside the family of birth; more precisely, this fosterage is a kind of return to the maternal family, since the child is brought up in the family of his father's sister – in reality in the home of his uncle on the mother's side, the sister's husband. It must not be forgotten that Polynesia is a region where there is a dual classification of kinship: maternal and masculine. Cf. our review of Elsdon Best's work, *Maori Nomenclature*, in *Année Sociologique* 7: 420, and Durkheim's observations in 5: 37.

12 Turner, *Nineteen Years in Polynesia*, p. 179; *Samoa*, p. 83.

13 Cf. our observations on *vasu* in Fiji, in 'Procès-verbaux de l'I.F.A', *Anthropologie*, 1921.

14 Krämer, *Samoa-Inseln*, see under: *toga*, vol. 1, p. 482; vol. 2, p. 90.

15 Ibid, vol. 2, p. 296; cf. p. 90 (*toga = Mitgift* ['dowry']); p. 94, exchange of the *oloa* against *toga*.

16 Ibid, vol. 1, p. 477. Violette, *Dictionnaire Samoan-Français*, under *toga*, expresses it well: 'riches of the region consisting of finely woven matting and *oloa*, riches such as houses, boats, cloth, and guns' (p. 194, col. 2); and he refers us back to *oa*, 'riches, possessions', which includes all foreign articles.

17 Turner, *Nineteen Years in Polynesia*, p. 179; cf. p. 186. Tregear, *Maori*

Comparative Dictionary, p. 468 (at the word *toga*, given under the heading *taonga*), muddles up the goods that bear this name and those that bear the name *oloa*. This is clearly a slip.

Rev. Ella, 'Polynesian Native Clothing', *JPS*, vol. 9, p. 165 describes the *ie tonga* ('mats') as follows:

> They were the main wealth of the natives; formerly they were used as a form of money in exchanges of property, at marriages and on occasions demanding special courtesy. They are often kept in the families as heir-looms (substitute goods), and many of the old *ie* are known and valued very highly as having belonged to some famous family.

Cf. Turner, *Samoa*, p. 120. All these expressions have their equivalent in Melanesia and North America, and in our own folklore, as we shall see.

18 Kramer, *Samoa-Inseln*, vol. 2, pp. 90, 93.

19 See Tregear, *Maori Comparative Dictionary*, under *taonga*: Tahitian, *tatoa*, 'to give property', *faataoa*, 'to compensate, to give property'; Marquises Islands, see Lesson, *Polynésiens*, vol. 2, p. 232, *taetae*; cf. Radiguet, *Derniers Sauvages*, *tiau tae-tae*, 'presents given, gifts and goods of their country given in order to obtain foreign goods'. The root of the word is *tahu*, etc.

20 See M. Mauss (1914), 'Origines de la notion de monnaie', *Anthropologie*, ('Procès-verbaux de l'I.F.A.'), in which almost all the facts cited, except those concerning Central Africa and America, relate to this area.

21 G. Gray, *Proverbs*, p. 103 (translation, p. 103).

22 C.O. Davis, *Maori Mementos*, p. 21.

23 In *Transactions of the New Zealand Institute*, vol. 1, p. 354.

24 Theoretically the tribes of New Zealand are divided, by Maori tradition itself, into fishermen, cultivators, and hunters, and are deemed to exchange their products with one another constantly. Cf. Elsdon Best, 'Forest Lore', *Transactions of the New Zealand Institute*, 42: 435.

25 Ibid, Maori text, p. 431, transl. p. 439.

26 The word *hau* designates, as does the Latin *spiritus*, both the wind and the soul – more precisely, at least in certain cases, the soul and the power in inanimate and vegetal things, the word *mana* being reserved for men and spirits. It is applied less frequently to things than in Melanesian.

27 The word *utu* is used for the satisfaction experienced by blood-

avengers, for compensations, repayments, responsibility, etc. It also designates the price. It is a complicated notion relating to morality, law, religion, and economics.

28 *He hau*. The whole translation of these two sentences has been short-ened by Elsdon Best, whom I am nevertheless following.

29 A large number of facts to illustrate this last point had been gathered by R. Hertz for one of the paragraphs of his translation of *Sin and Expiation*. They demonstrate that the punishment for theft is merely the magical and religious effect of *mana*, the power that the owner retains over the good that has been stolen. Moreover, the good itself, hedged in by taboos and marked with the signs of ownership, is completely charged by these with *hau*, spiritual power. It is this *hau* that avenges the person suffering the theft, which takes possession of the thief, casts a spell upon him, and leads him to death or obliges him to make restitution. These facts are to be found in the book by Hertz, which we shall be publishing, under the paragraphs relating to *hau*.

30 In R. Hertz's work are to be found the documents relating to the *maori* to which we refer here. These *maori* are at the same time talismans, palladiums, and sanctuaries in which dwells the spirit of the clan, *hapu*, its *mana*, and the *hau* of its soil.

The documents of Elsdon Best concerning this point require com-ment and discussion, in particular those that relate to the remarkable expressions of *hau whitia* and of *kai hau*. The main passages are in 'Spiritual Concepts', *Journal of the Polynesian Society* 10: 10 (Maori text); and 9: 198. We cannot deal with them as we should, but what follows is our interpretation: '*hau whitia*, averted *hau*', states Elsdon Best, and his translation seems exact. For the sin of theft or that of nonpayment or nonrendering of total counter-services is indeed a perverting of the soul, of *hau*, such as in cases (where it is confused with theft) of the refusal to enter into an exchange or give a present. On the contrary, *kai hau* is badly translated when it is considered as the mere equivalent of *hau whitia*. It does indeed designate 'the act of eating the soul' and is certainly the synonym of *whangai hau*: cf. Tregear, *Maori Comparative Dictionary* (*MCD.*), under the headings of *kai* and *whangai*; but this equivalence is not a simple one. For the typical present is that of food, *kai*, and the word refers to that system of food communion, and to the wrong that persists by remaining unredressed. There is something more: the word *hau* itself comes into the same order of ideas: Williams, *Maori Dictionary*, p. 23, under the heading *hau*, states, 'present given as a form of thanks for a present received'.

31 We draw attention also to the remarkable expression *kai-hau-tai*, Tregear, *MCD*, p. 116: 'to give a present of food offered by one tribe to another; "festivity" (South Island)'. The expression means that this present and the festivity returned are really the soul of the first 'service' returning to its point of departure: 'food that is the *hau* of the food'. In these institutions and these ideas are intermingled all sorts of principles between which our European vocabularies, on the contrary, take the greatest care to distinguish.

32 Indeed the *taonga* seem to be endowed with individuality, even beyond the *hau* that is conferred upon them through their relationship with their owner. They bear names. According to the best enumeration (that of Tregear, loc. cit., p. 360, under the heading *pounamu*, extracted from the Colenso manuscripts), they specifically include only the following categories: the *pounamu*, the famous jades, the sacred property of the chiefs and the clans, usually the *tiki*, very rare, very personal, and very well carved; then there are various sorts of mats, one of which, doubtless emblazoned as in Samoa, bears the name *korowai*. (This is the sole Maori word that evokes for us the Samoan word *oloa*, the Maori equivalent of which we have failed to discover.)

A Maori document gives the name of *taonga* to the *karakia*, the individually named magic formulas that are considered to be personal talismans capable of being passed on: *JPS* 9: 126 (transl. p. 133).

33 Elsdon Best, 'Forest Lore', p. 449.

34 Here might be placed the study of the system of facts that the Maoris class under the expressive term of 'scorn of *Tahu*'. The main document relating to this is to be found in Elsdon Best, 'Maori Mythology', in *JPS* 9: 113. *Tahu* is the 'emblematic' name for food generally; it is its personification. The expression *Kaua e tokahi ia Tahu* – 'do not scorn Tahu' is used for a person who has refused the food that has been put before him. But the study of these beliefs concerning food in Maori areas would carry us far. Suffice it to say that this god, this hypostasis of food, is identical with Rongo, the god of plants and peace. Thus we shall understand better the association of ideas between hospitality, food, communion, peace, exchange, and law.

35 See Elsdon Best, 'Spiritual Concepts', *JPS* 9: 198.

36 See Hardeland, *Dayak Wörterbuch*, vol. 1, pp. 190, 397a, under the headings *indjok, irak, pahuni*. The comparative study of these institu-

tions may be extended over the whole area of Malaysian, Indonesian, and Polynesian civilization. The sole difficulty consists in recognizing the institution. Let us give an example. It is under the heading of 'forced trade' that Spenser St John describes how, in the State of Brunei (Borneo), the nobles exacted tribute from the Bisayas by first making them gifts of cloth that were afterwards paid for at an usurious rate over a number of years (*Life in the Forests of the Far East*, vol. 2, p. 42). The error already arose among the civilized Malaysians themselves, who exploited a custom of their less civilized brothers, and no longer understood them. We shall not list all the Indonesian facts of this kind (see elsewhere the review of the study by M. Kruyt, *Koopen in Midden Celebes*).

37 To omit to invite someone to a war dance is a sin, a wrong that in the South Island bears the name of *puha*. See H.T. de Croisilles, 'Short Traditions of the South Island', *JPS* 10: 76 (note: *tahua*, 'gift of food').

The ritual of Maori hospitality includes: an obligatory invitation that the new arrival cannot refuse, but which he must not request either. He must make his way to the house of his host (who differs according to his caste) without looking about him. His host must have a meal prepared expressly for him, and must be humbly present. Upon leaving, the stranger receives a parting present (Tregear, *Maori Race*, p. 29). Cf. p. 1, the *identical* rites of Hindu hospitality.

38 In reality the two rules blend inextricably together, as do the antithetical and symmetrical services that they prescribe. A proverb expresses this intermingling: Taylor (*Te ika a maui*, p. 132, proverb no. 60) translates it roughly, 'When raw it is seen, when cooked, it is taken'. 'It is better to eat half-cooked food than to wait until the strangers have arrived', when it is cooked and one has to share it with them.

39 Chief Hekemaru (mistake for Maru), according to the legend, refused to accept 'the food' unless he had been seen and greeted by the village to which he was a stranger. If his retinue had passed by unnoticed and messengers had then been sent to request that he and his companions should retrace their steps and share in the eating of food, he would reply that 'the food should not follow after his back'. By this he meant that the food offered to 'the sacred back of his head' (namely, when he had gone beyond the village) would be dangerous for those who gave it to him. Hence the proverb, 'the food will not follow Hekemaru's back' (Tregear, *Maori Race*, p. 79).

40 The Tuhoe tribe commented upon these principles of mythology and

law to Elsdon Best ('Maori Mythology', *JPS* 8: 113). 'When a famous chief is to visit the locality, his *mana* precedes him.' The people in the area set out to hunt and fish in order to procure good food. They catch nothing: 'it is because our *mana* who has gone ahead' has made all the animals and fish invisible; 'our *mana* has banished them . . . ' etc. (There follows an explanation of the ice and snow, of the *Whai riri* [the sin against water], which keeps the food away from men). In reality this somewhat obscure commentary describes the state of a territory of a *hapu* of hunters whose members had not done what was necessary in order to receive the chief of another clan. They would have committed a '*kaipapa*, a sin against the food', and thus have destroyed their harvests, their game and fisheries, their own food.

41   Examples: the Arunta, the Unmatjera, and the Kaitish (cf. Spencer and Gillen, *Northern Tribes of Central Australia*, p. 610).

42   On the *vasu* see in particular the old treatise of Williams (1858) *Fiji and the Fijians*, vol. 1, p. 34. See also Steinmetz, *Entwicklung der Strafe*, vol. 2, p. 241 ff. This right of the nephew on the mother's side merely corresponds to the family communism system. But it allows one to gain some idea of other rights, for example, those of relations by marriage and what is generally called 'legal theft'.

43   See Bogoras, *The Chukchee* (Jesup North Pacific Expedition, Memorandum of the American Museum of Natural History), vol. 7, New York. The obligations to be carried out for receiving and reciprocating presents, and for hospitality, are more marked among the Chukchee of the maritime areas than among those living in reindeer country. Cf. *Social Organization* . . . , pp. 634, 637, Cf. the rule for the sacrifice and the slaughter of reindeer. Cf. *Religion* . . . , vol. 2, p. 375: the duty to invite, the right of the guest to ask for whatever he wants, and the obligation laid upon him to give a present.

44   The theme of the obligation to give is a profoundly Eskimo one. Cf. our study of the 'Variations saisonnières dans les sociétés eskimo' *Année Sociologique* 9: 121. One of the recent collections of stories of Eskimos published contains stories of this kind that preach generosity. Cf. Hawkes, *The Labrador Eskimos* (Canadian Geological Survey, Anthropological Series), p. 159.

45   We have (in 'Variations saisonnières dans les sociétés eskimo', *Année Sociologique* 9: 121) considered the festivities of the Alaskan Eskimos as a combination of Eskimo elements and of borrowings made from the Indian potlatch proper. But since writing about this, the potlatch, as well as the custom of presents, has been identified as existing

among the Chukchee and the Koryak of Siberia, as we shall see. Consequently the borrowing could just as well have been made from these as from the American Indians. Moreover, we must take into account the fine, and plausible hypotheses of Sauvageot (1924) (*Journal des Américanistes*) relating to the Asiatic origin of the Eskimo languages. These hypotheses are confirmed by the very strong ideas of archeologists and anthropologists about the origins of the Eskimos and their civilization. Finally, everything demonstrates that the Eskimos of the west, instead of being rather degenerate as compared with those of the east and the centre, are closer, linguistically and ethnologically, to the source. This seems now to have been proved by Thalbitzer.

In these conditions one must be more definite and say that potlatch exists among the eastern Eskimos and that it was established among them a very long time ago. However, there remain the totems and masks, which are somewhat peculiar to such festivals in the west, and a certain number of which are of Indian origin. Finally, the explanation is fairly unsatisfactory as accounting for the disappearance of the Eskimo potlatch from the east and centre of the American Arctic, unless it is explicable through the diminution in eastern Eskimo societies.

46   Hall, *Life with the Esquimaux*, vol. 2, p. 320. It is extremely remarkable that this expression has been given to us, not through observations made of the Alaskan potlatch, but as relating to the Eskimos of the centre, who only hold winter festivals for communistic activities and the exchange of presents. This demonstrates that the idea goes beyond the bounds of the institution of potlatch proper.

47   Nelson, 'Eskimos about Behring Straits', *Seventeenth Annual Report*, Bureau of American Ethnology, p. 303 ff.

48   Porter, *Alaskan Eleventh Census*, pp. 138, 141; and, especially, Wrangell, *Statische Ergebnisse . . .* , p. 132.

49   Nelson. Cf. 'asking stick' [sic] in Hawkes, *The Inviting-in Feast of the Alaskan Eskimos*, Geological Survey: Memoir 45, Anthropological Series 2, p. 7.

50   Hawkes, loc. cit, pp. 3, 7, 9 gives a description of one of these festivals, that of Unalaklit versus Malemiut. One of the most characteristic features of this collection is the comic series of 'total services' on the first day and the presents that they entail. The tribe that succeeds in making the other one laugh can ask from it what it likes. The best dancers receive valuable presents (pp. 12–14). It is a very clear and extremely rare example of ritual representations (I know of no other

examples save in Australia and America) of a theme which, on the contrary, is very frequent in mythology: that of the jealous spirit who, when he laughs, lets go of the thing that he is holding.

The rite of the 'Inviting-in festival' ends, moreover, by a visit from the *angekok* (*shamane*) to the *inua*, the spirit-men whose mask he wears, and who indicate to him that they have enjoyed the dances and will send him some game. Cf. the present made to the seals. Jenness (1922) 'Life of the Copper Eskimos', *Report of the Canadian Arctic Expedition*, vol. 12, p. 178, n. 2.

The other themes of the law of gifts are also very well developed. For example, the *näsnuk* chief has not the right to refuse any present, or dish presented, however rare it may be, under pain of being disgraced for ever. Hawkes, ibid, p. 9.

Hawkes is perfectly correct in considering (p. 19) that the festival of the Dene (Anvik) described by Chapman (1907) (*Congrès des Américanisles de Québec*, vol. 2) is a borrowing by the Indians from the Eskimos.

51 See figure in Bogoras, *The Chukchee*, vol. 7 (2): 403.

52 Bogoras, ibid, pp. 399–401.

53 Jochelson, 'The Koryak', *Jesup North Pacific Expedition*, vol. 6, p. 64.

54 Ibid, p. 90.

55 See p. 38, 'This for Thee'.

56 Bogoras, *The Chukchee*, p. 400.

57 On customs of this kind, see Frazer, *Golden Bough*, 8th edn, vol. 3, pp. 78–85, 91 ff.; vol. 10, p. 169 ff; vol. 5, pp. 1, 161.

58 On the Tlingit potlatch, see, pp. 38 and 41. This characteristic is basic to all the potlatches in the American Northwest. It is, however, hardly apparent because the ritual is too totemlike for its effect upon nature to be very marked, on top of its effect upon the spirits. In the Behring Straits area, particularly in the potlatch between the Chukchee and the Eskimos on St Lawrence Island, it is much more apparent.

59 See Bogoras, *Chuckchee Mythology*, p. 14, line 2 ff. for a potlatch myth. A dialogue is begun between two Shamans: 'What will you answer?' namely 'give as return present'. This dialogue finishes in a wrestling match. Then the two Shamans make a contract with each other. They exchange with each other their magic knife and their magic necklace, and their spirit (these attend upon magic), and finally their body (p. 15, line 2). But they are not perfectly successful in making their flights and landings. This is because they have forgotten to exchange their bracelets and their tassels, 'my guide in motion' (p. 16, line 10). In the end

they succeed in performing their tricks. It can be seen that all these things have the same spiritual value as the spirit itself, and are spirits.

60 See Jochelson, 'Koryak Religion', Jesup North Pacific Expedition, vol. 6, p. 30, A Kwakiutl chant of the dance of the spirits (the Shamanism of the winter ceremonies) comments upon the theme:

> You send us everything from the other world, O spirits, you who take away from men their senses.
>
> You have heard that we were hungry, O spirits . . .
>
> We shall receive much from you, etc . . .

See: Boas, *Secret Societies and Social Organization of the Kwakiutl Indians*, p. 483.

61 Davy, 'Foi jurée', p. 224, ff. See also p. 37.

62 *Koopen in midden Celebes*, Mededelingen der Koninglijke Akademie van Wetenschaapen, Afdeeling Letterkunde, 56; series B, no. 5, pp. 158, 159, 163–8.

63 Ibid, pp. 3, 5 of the extract.

64 Malinowski, *Argonauts of the Western Pacific*, p. 511.

65 Ibid, pp. 72, 184.

66 P. 512 (those who are not the objects of obligatory exchange). See Baloma (1917) 'Spirits of the Dead', *Journal of the Royal Anthropological Institute*.

67 A Maori myth, that of Te Kanava. Grey, *Polyn. Myth*, p. 213, tells how the spirits, the fairies, took on the shade of the *pounamu* (jades, etc.), alias *taonga*, laid out in their honour. Wyatt Gill, *Myths and Songs from the South Pacific*, p. 257 recounts an exactly identical myth from Mangaia, which tells the same story about necklaces made of discs of red mother-of-pearl, and how they won favour with the beautiful Manapa.

68 P. 513. Malinowski (*Argonauts of the Western Pacific*, p. 510 ff.) somewhat exaggerates the novelty of these facts, which are exactly identical to those of the Tlingit and Haïda potlatches.

69 'Het primitieve denken, voorn. in Pokkengebruiken', *Bijdr. tot de Taal-, Land- en Volksdenken v. Nederl, Indië*, vol. 71, pp. 245, 246.

70 Crawley, *Mystic Rose*, p. 386, has already launched a hypothesis of this kind and Westermarck has taken up the question and is beginning to prove it. See especially, *History of Human Marriage*, 2nd edn, vol. 1, p. 394 ff. But he did not see clearly its purport through not having identified the system of total services and the more developed system of potlatch in which all the exchanges, and particularly the exchange of

women and marriage, are only one of the parts. Concerning the fertility in marriage ensured by gifts made to the two spouses, see Ch. 3, n. 112, p. 152.

71 Vâjasaneyisamhita. See Hubert and Mauss, 'Essai sur le sacrifice', *Année Sociologique* 2: 105.

72 Tremearne (1913) *Haussa Superstitions and Customs*, p. 55.

73 Tremearne (1915) *The Ban of the Bori*, p. 239.

74 Robertson Smith, *Religion of the Semites*, p. 283. 'The poor are the guests of God.'

75 The Betsimisaraka of Madagascar tell of two chiefs, one of whom gave away everything that he possessed, while the other gave away nothing and kept everything for himself. God gave good fortune to the one who was generous, and ruined the miser (Grandidier, *Ethnographie de Madagascar*, vol. 2, p. 67.

76 On notions concerning alms, generosity, and liberality, see the collection of facts gathered by Westermarck, *Origin and Development of Moral Ideas*, vol. 1, chapter 23.

77 Concerning the value still attached at the present day to the magic of the *sadqâa*, see below.

78 We have not been able to carry out the task of re-reading an entire literature. There are questions that can only be posed after the research is over. Yet we do not doubt that by reconstituting the systems made up of unconnected facts given us by ethnographers, we would still find other important vestiges of the potlatch in Polynesia. For example, the festivals concerning the exhibiting of food, *hakari*, in Polynesia (see Tregear, *Maori Race*, p. 113) consist of exactly the same displays, the same heaps of food piled up one on another, the same distribution of food, as the *hakarai*, the same festivals with identical names among the Koita Melanesians. See Seligmann, *The Melanesians*, pp. 141–5, and *passim*. On the *hakari*, see also Taylor, *Te ika a Maoui*, p. 13; Yeats (1835) *An Account of New Zealand*, p. 139; Tregear, *Maori Comparative Dictionary*, under *hakari*. A myth in Grey, *Polyn. Myth*, p. 213 (1855 edn), and p. 189 (Routledge's popular edn), which describes the *hakari* of Maru, the god of war, in which the solemn designation of the recipients is absolutely identical to that in the festivals of New Caledonia, Fiji, and New Guinea. Below is also a speech constituting an *uma taonga* (*taonga* 'oven'), for a *hikairo* (food distribution), preserved in a song (given in Sir E. Grey (1835) *Ko nga Moteata: Mythology and Traditions in New Zealand*, p. 132), in so far as I am able to translate it (second verse):

Give me on this side my *taonga*,
Give me my *taonga*, so that I may heap them up,
That I may place them in a heap pointing towards land,
And in a heap pointing towards the sea,
Etc. . . . towards the east . . .
Give me my *taonga*.

The first verse doubtless refers to stone *taonga*. We can see just how much the very notion of the *taonga* is inherent in the ritual of the festival of food. See Percy Smith, 'Wars of the Northern against the Southern Tribes', *JPS* 8: 156 (the *hakari* of Te Toko).

79 Even assuming that the institution is not found in present-day Polynesian societies, it may well have existed in civilizations and societies that the immigration by Polynesians has absorbed or replaced, and it may well also be that the Polynesians had it before their migration. Indeed there is a reason for its having disappeared from part of this area. It is because the clans have definitively become hierarchized in almost all the islands and have even been concentrated around a monarchy. Thus there is missing one of the main conditions for the potlatch, namely the instability of a hierarchy that rivalry between chiefs has precisely the aim of temporarily stabilizing. Likewise, if we find more traces (perhaps of secondary origin) among the Maori than in any other island, it is precisely because chieftainship had been reconstituted there, and isolated clans had become rivals.

For the destruction of wealth on Melanesian or American lines in Samoa, see Krämer, *Samoa-Inseln*, vol. 1, p. 375. (See Index, under *ifoga*.) The Maori *muru*, the destructions of goods because of misdoing, may also be studied from this viewpoint. In Madagascar, the relations between the *Lohateny*, who should trade with one another, who may insult one another, and wreak havoc among themselves, are likewise vestiges of the ancient potlatches. See Grandidier, *Ethnographie de Madagascar*, vol. 2, p. 131 and n.; pp. 132–3. See also p. 155.

2 THE EXTENSION OF THIS SYSTEM: LIBERALITY, HONOUR, MONEY

*All these facts, like those that follow, are taken from fairly diverse ethnographical areas the connections between which it is not our purpose to study. From an ethnological viewpoint the existence of a Pacific civilization is unquestionable and partly explains many common features, for example, between the Melanesian and the American potlatch, as also the identity existing between the North Asian and the North American potlatch. However, on the other hand, the beginnings of potlatch that occurred among the Pygmies are very extraordinary. The traces of the Indo-European potlatch, which we shall also discuss, are no less so. We shall therefore refrain from taking into account all the fashionable considerations concerning the migrations of institutions. In our case it is too easy and too dangerous to speak of borrowing, and no less dangerous to speak of autonomous invention. Moreover, all the maps that are drawn up are no more than indications of our lack of knowledge or our present ignorance. For the time being, it must suffice for us to show the nature and very widespread diffusion of what is a legal theme. Let others write its history, if they can.

1 *Die Stellung der Pygmäenvölker*, 1910. We are not in agreement on this point with Fr Schmidt. See *Année Sociologique* 12: 65 ff.
2 *Andaman Islanders*, 1922, p. 83. 'Although the objects were regarded as presents they expected to receive something of equal value and were annoyed if the present returned did not come up to their expectations.'
3 Ibid, pp. 73, 81; see also p. 237. Brown then observes how unstable this state of contractual activity is, how it leads to sudden questioning, whereas often its aim is to prevent this.
4 P. 237.
5 P. 81.
6 The fact is indeed entirely comparable to the *kalduke* relationships of the *ngia-ngiampe*, among the Narrinyerri, and to the *Yutchin* among the Dieri. We reserve the right to return to these relationships.
7 P. 237.
8 Pp. 245–6. Brown puts forward an excellent sociological theory regarding these displays of communality, of identity of feelings, and of the obligatory and yet free character of their manifestations. Here there is another problem, which is moreover a related one, to which we have already drawn attention in 'Expression obligatoire des sentiments', *Journal de Psychologie*, 1921.
9 See Chapter 1, p. 18; Ch. 1, n. 79, p. 97.
10 There might be a need to take up once more the question of money, as regards Polynesia. See above, Chapter 1, n. 17, concerning the quotation from Ella regarding the Samoan mats. The large axes, the pieces

of jade, the *tiki*, the whale teeth – these are no doubt types of currency, as are also a large number of shells and crystals.

11 See 'La monnaie néo-calédonienne', *Revue d'Ethnographie*, p. 328 (1922), particularly regarding moneys circulating at the end of funeral rites, and the principle, p. 322. See also 'La fête du pilou en Nouvelle-Calédonie', *Anthropologie*, p. 226 ff.

12 Ibid, pp. 236–7; see also pp. 250, 251.

13 Ibid, p. 247; see also pp. 250–1.

14 'La fête du pilou . . . ', p. 263; 'La monnaie . . . ', p. 332.

15 This formula seems to belong to Polynesian juridical symbolism. In the Mangaia Islands peace was symbolized by a 'well-covered house' that brought together the gods and the clans underneath a 'well-woven roof': Wyatt Gill, *Myths and Songs of the South Pacific*, p. 294.

16 Fr Lambert (1900) in *Moeurs des Sauvages néo-calédoniens*, describes a number of potlatches: one in 1856, p. 119; the series of funeral festivals, pp. 234–5; a potlatch for a second burial, pp. 240–6. He has grasped the fact that the humiliation and even the emigration of a defeated chief was the sanction for a present and a potlatch that had not been reciprocated, p. 53; and he has understood that 'every present demands in return another present', p. 116; he uses the popular French expression *un retour* ('one back'), meaning a present reciprocated according to the rules; the *retours* are displayed in the chief's hut, p. 125. Presents at visits are obligatory. They are a condition of marriage, pp. 10, 93–4; they are irrevocable and 'return presents are made with interest', in particular to the *bengam*, the first cousin of a particular kind, p. 215. The *trianda*, the present dance (see p. 158), is a remarkable case in which formalism, ritualism, and juridical aestheticism are all mixed up together.

17 See '*Kula*', *Man*, July 1920, no. 51, p. 90 ff.; Malinowski (1922) *Argonauts of the Western Pacific*, London. All the references not otherwise indicated in this section refer to this book.

18 However, Malinowski exaggerates, on pp. 513 and 515, the novelty of the facts that he describes. First, the *kula* is in reality only an intertribal potlatch of a fairly common kind in Melanesia, and to which belong the expeditions described by Fr Lambert, and the great expeditions, such as the *Olo-Olo* of the Fijians, etc. See Mauss (1920) 'Extension du potlatch en Melanésie', ('Procès-verbaux de l'I.F.A.'), *Anthropologie*. The meaning of the word *kula* seems to me to be linked to that of other words of the same type, for example, *ulu-ulu*. See Rivers, *History*

of the Melanesian Society, vol. 1, p. 160; vol. 2, pp. 415, 485. However, in certain aspects even the *kula* is less characteristic than the American potlatch: the islands are smaller, the societies poorer and weaker than those of the coast of British Columbia. Among these latter all the features of intertribal potlatches are to be found. One even meets truly international potlatches, for example, Haïda against Tlingit (Sitka was in fact a town common to both of them, and the Nass River a place where they were constantly meeting); Kwakiutl against Bellacoola and Heiltauq; Haïda against Tsimshian, etc. This was moreover in the nature of things: the forms of exchange are normally capable of being extended and are international. Doubtless, there, as elsewhere, they have followed and pioneered commercial relations between these tribes, who were equally rich, and equally seafaring.

19 Malinowski favours the expression '*kula* ring'.

20 P. 97: 'noblesse oblige'.

21 See p. 473 for the expressions of modesty: 'the remnants of my food today, take them; I bring them to you', said whilst a precious necklace is being handed over.

22 See pp. 95, 189, 193. It is in a purely didactic way, and in order to make it understood by Europeans, that Malinowski (p. 187) places the *kula* among the 'ceremonial exchanges with payment' (in return): the words 'payment' and 'exchange' are both European.

23 See 'Primitive Economics of the Trobriand Islanders', *Economic Journal*, March, 1921.

24 The ritual of *tanarere*, the exhibition of the fruits of the expedition on the Muwa beach, pp. 374–5, 391. Cf. *Uvalaku* of Dobu, p. 381 (20–21 April). They decide who has been the most handsome, namely, the one who has been most fortunate and the best trader.

25 Ritual of the *wawoyla*, pp. 353–4; magic of the *wawoyla*, pp. 360–3.

26 See above, n. 21.

27 P. 471. See the frontispiece and the photographs of plates 60 ff. See p. 61.

28 Exceptionally, we shall here point out that one can compare these moral qualities with the fine paragraph in the *Ethics* of Nichomachus concerning the μεγαλοηβεπειχ and the ἐλευσεριχ.

29 A *Note* of principle concerning the use of the notion of money: in spite of the objections of Malinowski (1923) 'Primitive currency', *Economic Journal*, we continue to employ this term. Malinowski has previously protested against the abuse of the term (*Argonauts*, p. 499, n. 2), and has criticized Seligmann's nomenclature. He reserves the notion of

money for objects that not only serve as a means of exchange, but also as a standard to measure value. Simiand has made objections of the same kind to me concerning the use of the notion of value in societies of this kind. The two scholars are surely right, from their own viewpoint: they understand the word 'money' and the word 'value' in a restricted sense. On this reasoning there has only been economic value where there has been money, and there has only been money when precious things, themselves intrinsic forms of wealth and signs of riches, have been really made into currency, namely, have been inscribed and impersonalized, and detached from any relationship with any legal entity, whether collective or individual, other than the state that mints them. But the question posed in this way concerns only the arbitrary limit that must be placed on the use of the word. In my view, one only defines in this way a second type of money – our own.

In every society that has preceded those in which gold, bronze, and silver have been minted as money, there have been other things, stones, shells, and precious metals in particular, that have been used and have served as a means of exchange and payment. In a fair number of societies around us today this self-same system functions in reality, and it is this one we are describing.

It is true that these precious objects differ from what we are accustomed to conceive of as instruments for discharging debts. First, in addition to their economic nature and value, they have also a somewhat magic nature and are above all talismans – life-givers, as Rivers used to say, and as Perry and Jackson still do. They do indeed have very general circulation within a society and even between societies. But they are still attached to persons or clans (the first Roman currencies were minted by the *gentes*), to the individuality of their former owners, and to contracts drawn up between legal entities. Their value is still subjective and personal. For example, the money consisting of threaded shells in Melanesia is still measured in terms of the finger-span of the giver. (See Rivers, *History of the Melanesian Society*, vol. 1, pp. 64, 71, 101, 160 ff. vol. 2, p. 527. Cf. the expression *Schulterfaden* in Thurnwald, *Forschungen*, vol. 1, p. 189, v. 15; vol. 3, p. 41 ff. *Hüftschnur*, vol. 1, p. 263, line 6, etc. We shall see other important examples of these institutions. It is still true that these values are unstable, and lack that character of being a standard or measure. For example, their price rises or falls with the number and size of the transactions in which they have been used.

Malinowski very neatly compares with the jewels in a crown the *vaygu'a* of the Trobriand, which acquire prestige in the course of their voyages. Likewise, the emblazoned copper objects of the American Northwest and the mats of Samoa increase in value at each potlatch and in each exchange.

On the other hand, from two standpoints, these precious objects have the same function as money in our societies and consequently deserve at least to be placed in the same category. They have purchasing power, and this power has a figure set on it. For such and such an American copper object, a payment of so many blankets is due, to such and such *vaygu'a* correspond so many baskets of yams. The idea of number is present, even if that number is fixed in a different way from the authority of the state, and varies during the succession of *kula* and potlatches. Moreover, this purchasing power does indeed discharge debts. Even if it is recognized only between individuals, clans, and certain tribes, or only between associates, it is none the less public, official, and fixed. Brudo, Malinowski's friend, and like him long a resident of the Trobriand Islands, paid his pearl fishermen with *vaygu'a* as much as he did with European money or goods whose value was fixed. The passage from one system to another took place without a hitch, and was therefore possible. Armstrong, speaking about the moneys used on Rossel Island, close to the Trobriands, gives very clear indications of this, and, if there be error, persists in the same error as ourselves. (See 'A unique monetary system', *Economic Journal*, 1924, which we saw in proof.)

According to our view, for a long while humanity found its way was difficult. First, in the initial phase it found that certain things, almost all magic and precious, were not destroyed by use, and to these was given purchasing power (See Mauss (1914)) 'Origines de la notion de monnaie', *Anthropologie*, in 'Procès-verbaux de l'IFA'. At that time we had only discovered the early origin of money. Then, in a second phase, after having succeeded in putting these objects into circulation, within the tribe and in a wide area outside it, humanity found that these instruments of purchase could serve as a means of setting a figure on riches, and for putting these riches into circulation. This is the stage that we are describing. It is beyond this stage, in a fairly early era in Semitic societies, but doubtless not a very ancient one elsewhere, that – in a third phase – there was discovered the means of separating these precious objects from groups and people, in order to turn them into permanent instruments for the measure of value, and even a universal measure, although not a rational one – whilst waiting for something better to come along.

Thus there has been, in our opinion, a form of money that has preceded our own. This is without taking into account forms that consisted of useful objects; for example, in Africa and Asia, the sheets and ingots of copper, iron, etc., and without taking into account cattle in our ancient societies and in present-day African societies.

We apologise for having been obliged to take sides in these very far-ranging questions. But they touch very closely upon our subject, and it was necessary to be clear.

30 It seems that in the Trobriand Islands the women, like the princesses in the American Northwest, together with a few other persons, serve to some extent as a means of displaying the objects on show ... without taking into account the fact that they are 'enchanted' in this way. See Thurnwald, *Forsch. Salomo Inseln*, vol. 1, pp. 138, 159, 192, v. 7.

31 See below, n. 41.

32 Cf. 'Kula', *Man*, 1920, p. 101. Malinowski tells us that he has not found any mythical or other reasons for the direction this circulation takes. It would be very important to discover them. For, if there was any reason for the orientation of these objects, so that they tended to return to their point of origin, following a path of mythical origin, the fact would be miraculously identical to the Polynesian one, the Maori *hau*.

33 Regarding this civilization and trade, see Seligmann, *The Melanesians of British New Guinea*, Chapter 33 ff. See *Année Sociologique* 12: 374; *Argonauts*, p. 96.

34 The people of Dobu are 'hard on the *kula*' (*Argonauts*, p. 96).

35 Ibid.

36 Pp. 492, 502.

37 The 'remote partner' (*muri muri*, cf. *muri*, Seligmann, *Melanesians*, pp. 505, 752) is known at least by one section of the series of partners, like our modern banking correspondents.

38 See on pp. 89 and 90 the judicious observations of a general nature concerning ceremonial objects.

39 P. 504, names of pairs, pp. 89, 271. See the myth, p. 233: the way in which one hears a *soulava* spoken of.

40 P. 512.

41 P. 513.

42 P. 340; commentary, p. 341.

43 For the use of the horn-shaped shell, see pp. 340, 387, 471. Cf. Plate

61. This shell is the instrument that is sounded at every transaction, at every solemn moment during the common meal, etc. For the extension, if not the history, of the use of the shell, see Jackson (1921) *Pearls and Shells* (University of Manchester Series).

The use of trumpets and drums, as well as during festivals and contract ceremonies, is met with in a very large number of negro societies (Guinean and Bantu), Asian, American, and Indo-European etc. It is linked to the theme of law and the economy that we are studying here, and deserves a separate study in itself, and also for its history.

44 P. 340. *Mwanita, mwanita*. Cf. the text in *kiriwina* of the first two lines (the second and third, according to our reckoning), p. 448. This word is the name for long worms with black rings, with which are identified the necklaces made of discs of the spondylus, p. 341. There follows an evocation that is also an invocation:

> Come there together. I shall make you come there together. Come here together, I shall make you come here together. The rainbow appears there. I shall make the rainbow appear there. The rainbow appears here. I shall make the rainbow appear here.

Malinowski, according to the natives, considers the rainbow as a mere portent. But it may also designate the many variegated reflections given off by mother-of-pearl. The expression, 'Come here together', refers to the objects of value that will be brought together in the contract. The play on words between 'here' and 'there' is represented very simply by the sounds 'm' and 'w', which are a species of word-formers. They are very frequent in magic.

Then comes the second part of the exordium: 'I am the sole man, the sole chief, etc.' But this is only of interest from other viewpoints, in particular that of the potlatch.

45 The word that is translated like this (see p. 449) is *mwnumwaynise*, a double duplication of *mwana* or *mwayna*, which expresses 'itching' or 'state of excitement'.

46 I have presumed that there must be a line of this kind because Malinowski states categorically (p. 340) that this key word of the spell designates the state of mind that has possessed the partner, and which will make him give generous gifts.

47 The taboo was generally imposed because of the *kula* and the *s'oi*, funeral festivals, in order to gather together the food and palm nuts

that were necessary, as well as precious objects. See pp. 347, 350. The spell extends to the food.

48 Various names for necklaces. We do not analyse them in this study. These names are composites of *bagi*, necklace (p. 351), and various words. There are other special names for necklaces that are enchanted in the same way.

As this formula is one of the *kula* of Sinaketa, where necklaces are sought after and bracelets are neglected, only necklaces are mentioned. The same formula is used in the *kula* of Kiriwina. But then, since it is bracelets that are sought after, it would be the names of different kinds of bracelets that are mentioned, with the rest of the formula remaining the same.

The conclusion of the formula is also interesting, but again, only from the viewpoint of the potlatch:

> I am 'going *kula*' (to do my trading), I am going to deceive my *kula* (my partner). I am going to rob my *kula*, I am going to pillage my *kula*, I am going *kula* until my boat sinks . . . My fame is a clap of thunder. My tread, an earthquake.

The concluding line is strangely American in its outward form. There are analogous ones in the Solomon Islands. See Ch. 2, n. 257, p. 136.

49 P. 344; commentary, p. 345. The end of the formula is the same as that we have just quoted: 'I am "going *kula*", etc.'

50 P. 343. See p. 449, which gives the text of the first line with a grammatical commentary.

51 P. 348. This couplet comes after a series of lines (p. 347): 'Your fury, O man of Dobu, abates (as does the sea).' There then follows the same series with 'Woman of Dobu' (see p. 27). The women of Dobu are taboo, whereas those of Kiriwina prostitute themselves to visitors. The second part of the incantation is of the same kind.

52 Pp. 348, 349.

53 P. 356. Perhaps there is here an orientation myth.

54 Here one might use the term normally employed by Lévy-Bruhl: 'participation'. But the term has in fact its origin in confusion and muddle, and especially in legal identifications and communal procedures of the kind that we have now to describe.

Here we are dealing with the principle and it is unnecessary to go into the consequences.

55 P. 345 ff.

56 P. 98.

57 Perhaps there is also in this word an allusion to the ancient money that consisted of wild boar tusks.

58 Use of *lebu*, p. 318. See 'Myth', p. 313.

59 Violent denunciation (*injuria*), p. 357 (see numerous songs of this kind in Thurnwald, *Forsch.* vol. 1).

60 P. 359. It is said of a famous *vaygu'a*: 'Many men have died for him.' It seems, at least in one case, that of Dobu (p. 356), that the *yotile* is always a *mwali*, a bracelet, the feminine principle in a transaction: 'We do not *kwaypolu* or *pokala* them, they are women.' But in Dobu only bracelets are sought after and it may be that the phenomenon has no other meaning.

61 It appears that here several varied systems of transactions are mixed up together. The *basi* may be a necklace (see p. 98), or a bracelet of less value. But one can also give in *basi* other objects that are not strictly *kula*: spoons for lime (for the betel nut), crude necklaces, large polished axes (*beku*) (pp. 358, 481), which are also kinds of money, come into the process here.

62 Pp. 157, 359.

63 Malinowski's book, like that of Thurnwald, shows the superior observation of a true sociologist. Moreover, it was the observations of Thurnwald concerning the *mamoko* (vol. 3, p. 40), the *Trostgabe*, made to Buin that put us on the track of some of these facts.

64 P. 211.

65 P. 189. See Plate 37. See p. 100, 'secondary trade'.

66 See p. 93.

67 It seems that these gifts carry the generic name '*wawoyla*' (pp. 353–4; see pp. 360–1). See *Woyla*, '*kula* courting', p. 439, in a magic formula in which in fact are enumerated all the objects that the future partner can possess, and on whose 'ebullition' the donor must decide. Among these objects is in fact the series of presents that follows.

68 This is the most general term: 'presentation goods' (pp. 205, 350, 439). The word *vata'i* is the one that designates the same presents made by the people of Dobu. See p. 391. These 'arrival gifts' are enumerated in the formula: 'my lime pot is boiling; my spoon is boiling, my little basket is boiling, etc.' (The same theme and the same expressions. See p. 200.)

Besides these generic names, there are special names for various presents in various circumstances. The offerings of food that the Sinaketa people bring to Dobu (and not vice versa) bear the simple name of *pokala*, which corresponds fairly well to 'wages', 'offering', etc. Also termed *pokala* are the *gugu'a*, 'personal belongings' (p. 501; see also pp. 270, 313), which the individual deprives himself of in order to attempt to attract (*pokapokala*, p. 360) his future partner (see p. 369). In these societies there is a very strong sense of the difference existing between objects for personal use and those that are 'properties', durable things in the family and the circulation of goods.

69 E.g. p. 313, *buna*.

70 E.g. the *kaributu* (pp. 344, 358).

71 They say to Malinowski: 'My partner is the same thing as my fellow tribesman (*kakaveyogu*). He might fight against me. My true relative (*veyogu*), the same thing as an umbilical cord, would always be on my side' (p. 276).

72 This is what expresses the magic of the *kula*, the *mwasila*. See pp. 74–5.

73 The chiefs of the expedition and the chiefs of the boats have, in effect, precedence.

74 An amusing myth, that of the Kasabwaybwayreta (p. 342), brings together all these reasons. One learns how the hero obtains the famous necklace, Gumakarakedakeda, and how he outdistanced all his fellows in the *kula*. See also the myth of Takasikuna (p. 307).

75 P. 390. At Dobu, see pp. 362, 365.

76 At Sinaketa, not Dobu.

77 On trading in stone axes, see Seligmann, *Melanesians*, pp. 350, 353. The *korotumna* (*Argonauts*, pp. 358, 365) are normally decorated spoons of whalebone, decorated ladles that also serve as *basi*. There are still other intermediary gifts.

78 *Doga, dogina*.

79 Pp. 486–91. On the extension of these customs to all the civilizations designated as those of North Massim, see Seligmann, *Melanesians*, p. 584. There is a description of the *walaga* pp. 594, 603; see *Argonauts*, pp. 486–7).

80 P. 479.

81 P. 472.

82 The making and the gift of the *mwali* by brothers-in-law bears the name of *youlo* (pp. 280, 503).

83 P. 171 ff.; see also p. 98 ff.

84 For example, in the construction of boats, the collecting of items of pottery, or the provision of food.

85 The whole of tribal life is nothing more than a constant 'giving and

receiving'; every ceremony, every legal or customary act is carried out only with a material gift and a gift in return accompanying them. Wealth given and received is one of the principal instruments of social organization, of the chief's power, of the bonds of kinship through blood or marriage. (p. 167)

See pp. 175–6 and *passim* (see index: *Give and Take*).

86　It is often identical to that of the *kula*, with the partners often being the same (p. 193); for the description of the *wasi*, see pp. 187–8 and Plate 36.

87　The obligation still holds good today, in spite of the disadvantages and losses that the pearl fishermen suffer, obliged to carry on fishing and to lose considerable sums in wages to fulfil a purely social obligation.

88　See Plates 32, 33.

89　The word *sagali* means 'distribution' (as does the Polynesian word *hakari*), p. 491. The description is on pp. 147–50, 170, 182–3.

90　See p. 491.

91　This is especially evident in the case of funeral festivals. See Seligmann, *Melanesians*, pp. 594–603.

92　P. 175.

93　P. 323. Another term is *kwaypolu* (p. 356).

94　Pp. 354, 378–9.

95　Pp. 163, 373. The *vakapula* has subdivisions with special titles. For example, *vewoulo* (initial gift) and *yomelu* (final gift). (This proves the identity with the *kula*. Cf. the relationship *yotile vaga*.) A certain number of these payments have special names: *karibudaboda* designates the remuneration given to those who work on the boats and, in general, to those who work, for example, in the fields, and particularly for the final payments for crops (*urigubu*, in the case of annual services in gathering crops by a brother-in-law, pp. 63–5, 181) and for finishing off the making of necklaces (pp. 183, 394). It also bears the name of *sousala* when it is a large enough payment (making of discs of Kaloma, pp. 183, 373). *Youlu* is the name of the payment for making a bracelet. *Puwaya* is that of the food given as encouragement to the team of woodcutters. See the pretty song on p. 129;

> The pig, the coconut (drink) and the yams
> Are finished, and we are still dragging very heavy things along.

96　The two words *vakapula* and *mapula* are different moods of the verb *pula, vaka* being evidently the formative part of the causative. Concerning *mapula*, see pp. 178 ff., 182 ff. Malinowski often translates it by 'payment'. It is generally compared to a 'plaster', for it assuages the difficulty and fatigue arising from the service rendered, compensates for the loss of the object or secret given, and of the title and privilege that has been given away.

97 P. 179. The name of the 'gifts for sexual reasons' is also *buwana* and *sebuwana*.

98 See preceding notes: likewise *Kabigidoya* (p. 164) designates the ceremony of the presentation of a new boat, the people that made it, and the action they carry out ('to break the head of the new boat, etc.') ... and the gifts that, moreover, are returned with interest. Other words designate the location of the boat (p. 186); welcoming gifts (p. 232), etc.

99 *Buna*, gifts of 'big cowrie shell' (p. 317).

100 *Youlo, vaygu'a* given as reward for work on a crop (p. 280).

101 Pp. 186, 426, etc. designate evidently every usurious counter-service. For there is another name, *ula-ula*, for simple purchases by magic formulas (*sousala* when the prizes-cum-presents are very considerable – p. 183). *Ula'ula* is also said when the presents are offered to the dead as much as to the living (p. 183), etc., etc.

102 Brewster (1922) *Hill Tribes of Fiji*, pp. 91–2.

103 Ibid, p. 191.

104 Ibid, p. 23. One recognizes the word taboo, *tambu*.

105 Ibid, p. 24.

106 Ibid, p. 26.

107 Seligmann, *Melanesians* (Glossary, p. 754; and pp. 77, 93, 94, 109, 204).

108 See the description of the *doa*, ibid, pp. 71, 89, 91, etc.

109 Ibid, pp. 95, 146.

110 Kinds of money are not the only things in this system of gifts that these tribes of the Gulf of New Guinea call by a name identical to the Polynesian word and bearing the same meaning. We have already pointed out on Ch. 1, n. 78, p. 96, n. 65, the identical nature of the New Zealand *hakari* and the *hekarai*, festival exhibitions of food that Seligmann has described for us in New Guinea (Motu and Koita). (Cf. *Melanesians*, pp. 144–5, Plates 16–18).

111 See Ch. 1, n. 10, p. 88. It is remarkable that the word *tun*, in the Motu dialect (Banks Islands) – evidently identical to *taonga* – has the meaning of to buy (especially a woman). Codrington, in the myth of

Qat buying the night (*Melanesian Languages*, pp. 307–8, n. 9) trans- lates it: 'to buy at a great price'. In reality it is a purchase made according to the rules of potlatch, well attested to in this part of Melanesia.

112 See documents cited in *Année sociologique*, 12: 372.

113 See especially *Forsch.*, vol. 3, pp. 38–41.

114 *Zeitschrift fur Ethnologic*, 1922.

115 *Forsch.*, vol. 3, Plate 2, n. 3.

116 *In Primitive New Guinea*, 1924, p. 294.

117 Holmes in fact describes fairly inadequately the system of inter- mediary gifts.

118 See the study cited on in Ch. 1, n. 62, p. 95. The uncertain meaning of the words that we translate inadequately as 'to sell' and 'to buy' is not peculiar to Pacific societies. We shall come back later (Ch. 3, n. 106, p. 151,) to this subject, but already we would remind readers that even in our everyday language the word 'sale' can mean both sale and purchase, and that in Chinese there is only a tonal difference between the monosyllables that designate both the acts of selling and buying.

119 Since the eighteenth century, with the Russians, and since the beginning of the nineteenth century, with French Canadian trappers.

120 See, however, the sales of slaves: Swanton, 'Haïda Texts and Myths', *Bur. Am. Ethn. Bull.* 29: 410.

121 A summary bibliography of the theoretical works dealing with this potlatch is given on Intro., n. 6, p. 85; Intro, n. 13, p. 86.

122 This summary outline is presented without justification, but it is necessary. We warn that it is incomplete even as to the number and names of the tribes, and to their institutions.

We leave out a large number of tribes, principally the following: [1] Nootka (Wakash or Kwakiutl group), and the Bella Kula, who are neighbours; [2] the Salish tribes of the south coast. Moreover, research into the extension of the potlatch should be directed farther south, as far as California. There (and this is remarkable from other perspectives), the institution seems widespread in societies belong- ing to the so-called Penutia and Hoka groups. See, for example, Powers, 'Tribes of California', *Contributions to North American Eth- nology* 3: 153 (Pomo), 238 (Wintun), 303, 311 (Maidu); for other tribes see pp. 247, 325, 332, 333; for general remarks see p. 411.

Finally, the institutions and the arts we describe in a few words are infinitely complicated, and certain ones that are omitted are no less curious than those that are present. For example, pottery is unknown there, as it is in the lowest layer of the civilization of the South Pacific.

123   The sources that allow us to study these societies are considerable. They are remarkably reliable, being abundantly philological in content, and made up of texts that have been transcribed and translated. See summary bibliography in Davy, 'Foi jurée', pp. 21, 171, 215. To this should be principally added: F. Boas and G. Hunt (1921) 'Ethnology of the Kwakiutl' (henceforth *Ethn. Kwa.*), *35th Annual Report of the Bureau of American Ethnology*; F. Boas (1916) 'Tsimshian Mythology', *31st Annual Report of the Bureau of American Ethnology*, published 1923, henceforth, *Tsim. Myth.*). However, all these sources have one disadvantage: either the earlier ones are defective, or the new ones, despite their detail and depth, are not complete enough for our present concern. The attention of Boas and his companions on the Jesup Expedition was focused on the material aspects of civilization, on linguistics and mythological literature. Even the studies by the earlier professional ethnographers such as Krause and Jacobsen, or the more recent ones by Sapir, Hill Tout, etc. follow along the same lines. The juridical and economic analysis, as well as the demography, either remains to be carried out or at least to be supplemented. (However, the social morphology is commented upon in the various censuses of Alaska and British Columbia). Barbeau promises us a complete monograph on the Tsimshian. We await that indispensable information and we would like to see his example imitated very shortly, before time runs out. On numerous points concerning the economy and law, the old documents are the best: those of Russian travellers, those of Krause (*Tlinkit Indianer*) and Dawson (on the Haïda, the Kwakiutl, the Bellakoula, etc.), most of which have appeared in the *Bulletin* of the Geological Survey of Canada or in the *Proceedings* of the Royal Society of Canada; those of Swan (1870) (Nootka), 'Indians of Cape Flattery', *Smiths. Contrib. to Knowledge*; those of Mayne (1862) *Four years in British Columbia*, London. The dates of these works impart definitive authority to them.

In the nomenclature of these tribes, a difficulty arises. The Kwakiutl form a tribe and also give their name to several other tribes that, confederated with them, form a real nation under that name. Each time we shall try to make it clear to which Kwakiutl tribe we are referring. When no further details are given, the reference will be to the Kwakiutl proper. The word Kwakiutl, moreover, merely means rich, 'the smoke of the world', and itself points out the importance of the economic facts we are about to describe. We shall not reproduce here all the spelling details of these languages.

124 On the blankets of Chilkat, see Emmons, 'The Chilkat Blanket', *Mem. of the Amer. Mus. of Nat. History*, vol. 3.

125 See Rivet, in Meillet and Cohen, *Langues du Monde*, p. 616 ff. It was Sapir (1915), 'Na-Déné Languages', *American Anthropologist*, who definitively reduced the Tlingit and Haïda languages to branches from an Athapascan root.

126 On such payments for the acquisition of ranks, see Davy, 'Foi jurée', pp. 300–5. For Melanesia, see examples in Codrington, *Melanesians*, p. 106 ff.; Rivers, *History of the Melanesian Society*, vol. 1, p. 70 ff.

127 This word 'ascension' must be taken both literally and figuratively. Just as the ritual of the *vājapeya* (post-Vedic) contains a ritual of the ascension of a ladder, so the Melanesian ritual consists in having the young chief mount a platform. The Snahnaimuq and the Shushwap of the Northwest also have a stand from which a chief distributes his potlatch. Boas (1891) 'Fifth Report on the Tribes of North-Western Canada', *British Association for the Advancement of Science*, p. 39; (1894) 'Ninth Report', *Br. Ass.*, p. 459. The other tribes are only familiar with the platform on which sit the chiefs and the higher brotherhoods.

128 It is in this way that the older authors – Mayne, Dawson, Krause, etc. – describe its mechanism. See especially Krause, *Tlinkit Indianer*, p. 187 ff., which is a collection of documents of older authors.

129 If the hypothesis of the linguists is exact and if the Tlingit and the Haïda are merely Athapascans who have adopted the civilization of the Northwest (a hypothesis that is not far from that of Boas), the primitive character of the Tlingit and Haïda potlatch would be self-explanatory. It is also possible that the violence of the potlatch in the American Northwest arises from the fact that this civilization is at the meeting point of two groups of families of peoples who both enjoyed this civilization, with one form of civilization coming from southern California and the other from Asia.

130 Davy, 'Foi jurée', p. 247 ff.

131 Boas (1898) has written nothing better on the potlatch than the following pages drawn from the 'Twelfth Report on the North-Western Tribes of Canada', *Br. Ass.* pp. 54–5 (cf. Boas (1891) 'Fifth Report', p. 38):

The economic system of the Indians of the British colony is largely based on credit, as much as that of civilized peoples. In all his undertakings the Indian trusts to the aid of his friends. He promises to pay them for this assistance at a later date. If the aid provided consists of valuable things, which are measured by the Indians in blankets, just as we measure them in money, he promises to pay back the value of the loan with interest. The Indian has no system of writing and consequently, to guarantee the transaction, the promise is made in public. To contract debts on the one hand, to pay them on the other, this constitutes the potlatch. This economic system is developed to such an extent that the capital possessed by all the individuals associated with the tribe far exceeds the quantity of available valuables that exists; in other words, the conditions are entirely analogous to that prevailing in our own society: if we desired to pay off all our debts, we would find that there was not nearly enough money, in fact, to settle them. The result of an attempt by all creditors to seek reimbursement of their loans is a disastrous panic that the community takes a long time to recover from.

One must indeed understand that an Indian who invites all his friends and neighbours to a great potlatch, which apparently squanders all the profits accumulated over long years of work, has two things in view that we can only acknowledge to be wise and praiseworthy. His first purpose is to pay off his debts. This is done publicly with much ceremony, and is like a notarial act. His second purpose is to place the fruits of his labour so that he draws the greatest benefit from them for himself as well as for his children. Those who receive presents at this festival, receive them as loans that they use in ongoing enterprises, but after a few years they must be given back with interest to the donor or his heir. Thus the potlatch ends up by being considered by the Indians as a means of ensuring the wellbeing of their children, if they are left as orphans when they are young.

By correcting the terms 'debt', 'payment', 'reimbursement', 'loan', and replacing them with such terms as 'presents given' and 'presents returned', terms that Boas moreover ends up by using himself, we have a fairly exact idea of how the notion of credit functions in the potlatch.

On the notion of honour, see Boas, 'Seventh Report on the North-Western Tribes', p. 57.

132 A Tlingit expression: Swanton, *Tlingit Indians*, p. 421.

133 We have failed to notice that the notion of 'term' [of a debt] was not

only as ancient, but as simple, or, if you wish, as complex as the notion of cash.

134　Cuq (1910) 'Etude sur les contrats de l'époque de la première dynastie babylonienne', *Nouv. Rev. Hist. du Droit*, p. 477.

135　Davy, 'Foi jurée', p. 207.

136　Distribution of all the property: Kwakiutl, see Boas (1895) 'Secret Societies and Social Organization of the Kwakiutl Indians', *Rep. Amer. Nat. Mus.* (henceforth, *Sec. Soc.*), p. 469. In the case of the initiation of a novice, ibid, p. 551, Koskimo, Shushwap: redistribution, Boas (1890) 'Seventh Report', p. 91. Swanton, 'Tlingit Indians', 21st Annual Report, *Bur. of Am. Ethn.* (henceforth, *Tlingit*), p. 442 (in a speech): 'He has spent everything in order to put him (his nephew) on display.' Redistribution of everything that has been won by gambling: see Swanton, 'Texts and Myths of the Tlingit Indians', *Bulletin* no. 39 of *Bur. of Am. Ethn.* (henceforth *Tlingit T.M.*) p. 139.

137　On the war of property, see the song of Maa, *Sec. Soc.*, pp. 577, 602: 'We fight against property.' The opposition, the war of wealth, the war of blood, is to be found in the speeches made at the same potlatch of 1895 at Fort Rupert. See Boas and Hunt, *Kwakiutl Texts*, first series; Jesup Expedition, vol. 3 (henceforth, *Kwa.*, vol. 3), pp. 482, 485; see also *Sec. Soc*, pp. 668, 673.

138　See especially the myth of Haïyas (*Haïda Texts*, Jesup 6 (83), Masset) who has lost face in a gambling game and dies in mortification. His sisters and his nephews go into mourning, give a 'revenge' potlatch, and he comes to life again.

139　On this subject it would be necessary to study gambling, which even in French society is not considered to be a contract, but a situation in which honour is committed and where goods are handed over that, after all, one could refuse to hand over. Gambling is a form of potlatch, and of the gift system. Its extension, even as far as the American Northwest, is remarkable. Although the Kwakiutl knew of it (see *Ethn. Kwa.*, p. 1394, under the heading *ebayu*, dice (?), under the heading *lepa*, p. 1435, and compare *lep*, p. 1448, 'Second potlatch, dance'; see p. 1423, under the heading *maqwacte*), it does not appear to play a role among them comparable to that among the Haïda, Tlingit, and Tsimshian. The latter are hardened, incessant gamblers. See the descriptions of the game of 'tip and run' among the Haïda: Swanton, *Haïda* (Jesup Exp., vol. 1), pp. 58 ff.; 141 ff., for the faces and the names; the same game exists among the Tlingit; a description with the names of the sticks: Swanton, *Tlingit*, p. 443. The *naq*, the ordinary *tlingit*, the piece that wins, is equivalent to the Haïda *djil*.

Stories are full of legends about games, of chiefs that have lost everything through gambling. One Tsimshian chief has even lost his children and his parents: *Tsim. Myth.*, pp. 101, 207, cf. Boas, ibid, p. 409. A Haïda legend tells the story of an all-out game of the Tsimshians against the Haïda. See *Haïda T.M.*, p. 322. Cf. the same legend: the games against Tlingit, ibid, p. 94. One can find a list of themes of this kind in Boas, *Tsim. Myth.*, pp. 843, 847. Etiquette and morality decree that the winner should leave the loser, his wife, and his children free. See *Tlingit T.M.*, p. 137. There is no need to point out the connection between this feature and the Asiatic legends.

Moreover, there are irrefutable Asiatic influences here. On the spread of Asiatic games of chance to America, see the fine study by E.R. Tylor (1896), 'On American Lot-Games as Evidence of Asiatic Intercourse' (Bastian, *Festschrift*), in Supplement to *Int. Arch. J. Ethn*, p. 55 ff.

140 Davy has expounded the theme of the challenge, and rivalry. To this must be added that of the wager. See, for example, Boas, *Indianische Sagen*, pp. 203–6. There are wagers concerning food, wrestling, and 'ascension', etc. in the legends. See ibid, p. 363 for a list of themes. The wager is still nowadays a remnant of these rights and this morality. It commits only one's honour and credit, yet causes wealth to circulate.

141 On the potlatch of destruction, see Davy, 'Foi jurée', p. 224. To his account must be added the following remarks. To give is already to destroy (see *Sec. Soc.*, p. 334). A certain number of rituals of giving entail acts of destruction, for example, the ritual of the reimbursement of the dowry, or, as Boas calls it, 'repayment of the marriage debt', includes a formality termed 'sinking the boat' (see *Sec. Soc.* pp. 518, 520). But this ceremony is figurative. However, the visits to the Haïda or Tsimshian potlatch carry with them the real destruction of the boats of the arriving tribe. With the Tsimshian they are destroyed upon arrival, after help has been carefully given in unloading all they contain, and finer boats are handed over on departure (see Boas, *Tsim. Myth.*, p. 318).

But the act of destruction proper seems to constitute a higher form of extravagance. It is called 'killing property' among the Tsimshian and the Tlingit (see Boas, *Tsim. Myth.*, p. 344; Swanton, *Tlingit*, p. 442). In reality this term is even applied to the distribution of blankets: 'so many blankets were destroyed for appearances sake', *Tlingit*, ibid.

In this practice of destruction at the potlatch two other motives come into play: [1] the theme of war: the potlatch is a war. Among the Tlingit it bears the name of 'War Dance' (Swanton, *Tlingit*, pp. 436, 458). In the same way as in a war one can take possession of the masks, the names, and the privileges of their owners who have been killed, so in a war between properties, property is killed – either one's own, so that others cannot have it, or that of others, by giving them goods that they will be obliged to reciprocate, or will not be able to do so. [2] The second theme is that of sacrifice (see p. 15). If one kills property, it is because it has life (see p. 44). A herald says: 'May our property remain alive through the efforts of our chief, let our copper object not be broken' (*Ethn. Kwa.*, p. 1285, line 1). Perhaps even the meanings of the word *yäq*, 'to be laid out dead, to distribute a potlatch' are explicable in this way (see *Kwa.*, vol. 3, p. 59, line 3, and Index, *Ethn. Kwa.*).

However, in principle, as in normal sacrifice, it is a matter of passing on the things destroyed to the spirits, namely, to the ancestors of the clan. This theme is naturally more developed among the Tlingit (Swanton, *Tlingit*, pp. 443, 462), where the ancestors are not only present at the potlatch and benefit from the destructions, but even profit from the presents that are given their living homonyms. Destruction by fire seems to be a characteristic of this theme. Among the Tlingit see the very interesting myth in *Tlingit T.M*, p. 82.; Haïda sacrifice in the fire (Skidegate), Swanton, 'Haïda Texts and Myths', *Bull. Bur. Amer. Ethn.* (29) (henceforth *Haïda T.M.*), pp. 28, 36, 91. The theme is less prominent among the Kwakiutl, who, however, have a god called, 'Seated on the fire', and to whom, for example, the clothing of a sick child is sacrificed in order to pay off the divinity (see *Ethn. Kwa.*, pp. 705, 706).

142 Boas, *Sec. Soc.*, p. 353, etc.

143 See below, n. 209, concerning the word *p!Es* (sic).

144 It seems that even the words 'exchange' and 'sale' are foreign to the Kwakiutl language. I can only find the word 'sale' in the various glossaries of Boas dealing with the putting up for sale of a copper object. But this putting up for auction is nothing like a sale. It is a kind of wager, a struggle in generosity. As for the word 'exchange', I can only find it under the form *L'ay*: but in the text to which it refers (in *Kwa.*, vol. 3, p. 77, line 41), it is used in relation to a change of name.

145 See the expression, 'avid for food', *Ethn. Kwa.*, p. 1462, 'Desirous of making his fortune quickly'; ibid, p. 1394; see the fine imprecation against the 'little chiefs': 'The little men who deliberate; the little men who work; . . . who are conquered; . . . who promise to give boats; . . . who accept the property given; . . . who seek after property; . . . who work only for property' (the word that translates as property is *maneq*, 'to pay back a favour', ibid, p. 1403, 'the traitors', ibid, p. 1287, lines 15–18. See also another speech where, it is said of the chief who has given potlatch and of those people who receive and never give in return: 'He gave them to eat, he caused them to come . . . he took them on his own back', ibid, p. 1293; see also p. 1291, and another imprecation against the 'little men', ibid, p. 1381.

It must not be thought that a morality of this kind is contrary to the economic system or corresponds to a communistic kind of laziness. The Tsimshian blame avarice and tell how their main hero, Crow (the creator), was dismissed by his father because he was miserly: *Tsim. Myth.*, pp. 61, 444. The same myth exists among the Tlingit. The latter also blame the laziness and the begging nature of their guests and tell how Crow and those who go from town to town cadging invitations were punished: *Tlingit T.M.*, pp. 260, 217.

146 '*Injuria*: Mélanges Appleton'; 'Magie et droit individuel', *Année Sociologique* 10: 28.

147 Among the Tlingit one pays for the honour of dancing: *Tlingit T.M.*, p. 141. There is payment of the chief who composed the dance. Among the Tsimshian: 'Everything is done for honour . . . Above everything is wealth and the display of vanity', Boas (1889) 'Fifth Report', p. 19. Duncan, in Mayne, *Four years*, p. 265, had already said: 'for the sheer vanity of the thing'. Moreover, a large number of the rituals, not only that of the 'ascension', etc., but even those that, for example, consist of 'lifting the copper object' (Kwakiutl), (*Kwa.*, vol. 3, p. 409, line 26), 'lifting the spear' (Tlingit) (*Tlingit T.M.* p. 117), 'lifting the post of the potlatch', whether the funeral or totem post, 'lifting the centre-post of the home', the old slippery pole – all translate such principles into reality. We must not forget that the purpose of the potlatch is to know which is 'the most "exalted" family'. (For the commentaries of Chief Katishan on the myth of the Crow [Tlingit], see *Tlingit T.M.*, p. 119, note (a).

148 Tregear, *Maori Comparative Dictionary*, under the heading of *Mena*.

There would be a place for studying the notion of wealth itself. From our point of view, the rich man is one who has *mana* in

Polynesia, *auctoritas* in Rome, and who, in these American tribes is 'open-handed', *walas* (*Ethn. Kwa.*, p. 1396). But strictly speaking we need only point out the relationship between the notion of wealth, that of authority, and the right of commanding those who receive presents, and the potlatch: it is a very clear relationship. For example, among the Kwakiutl, one of the most important clans is that of the Walasaka (which is at the same time the name of a family, a dance and a brotherhood). The name means 'the great ones who come from on high', who distribute at a potlatch. *Walasila* means not only wealth, but also 'distribution of blankets on the occasion of the auctioning off of a copper object'. Another metaphor consists in considering that the individual is made 'heavy' by the potlatches that have been given (see *Sec. Soc.*, pp. 558, 559). The chief is said to 'swallow the tribes' to whom he distributes his wealth; he 'spits forth property', etc.

149 A Tlingit song says about the phratry of the Crow: 'It is that which makes the Wolves "valuable"', *Tlingit T.M.*, p. 398, no. 38. The principle that the 'respects' and 'honours' to be paid and received includes the gifts is very clear-cut in the two tribes – Swanton, *Tlingit*, p. 451. Swanton, *Haïda*, p. 162 shows that the need to return certain presents may be dispensed with.

150 See Conclusion, n. 8, p. 155.

The etiquette of the feast, of the gift that one receives with dignity, but is not solicited, is extremely marked among these tribes. Let us merely point to three facts relating to the Kwakiutl, Haïda, and Tsimshian practices that are instructive from our viewpoint: the chiefs and the nobles eat little at the feast; it is the vassals and the common people who eat a lot; the former literally purse their lips: Boas, *Kwa. Ind.*, Jesup Expedition, vol. 2, pp. 427, 430; on the dangers of eating too much, *Tsim. Myth.*, pp. 59, 149, 153, etc. (myths); they sing at the feast; *Kwa Ind.*, Jesup Expedition, vol. 2, pp. 430, 437. The seashell is sounded, 'so that no one should say we are dying of hunger', *Kwa.*, vol. 3, p. 486. The noble never requests anything. The Shaman doctor never asks for payment. His 'spirit' forbids him to do so (*Ethn. Kwa.*, pp. 731, 742; *Haïda T.M.*, pp. 238, 239). However, there exists among the Kwakiutl a brotherhood and a 'begging' dance.

151 See Introduction, n. 14, p. 87.

152 The Tlingit and Haïda potlatches have especially developed this principle. Cf. *Tlingit Indians*, p. 443, 462; cf. speech in *Tlingit T.M.*, p. 373; the spirits smoke while the guests are smoking; cf. p. 385, line 9:

'We who dance here for you, we are not really ourselves. It is our uncles who died long ago who are dancing here.' The guests are spirits, good-luck talismans *gona' gadet*, ibid, p. 119, note (a). In fact we have here purely and simply the confusion of the two principles of the sacrifice and the gift, comparable, except perhaps for its effect upon nature, to all the cases we have already cited above. To give to the living is to give to the dead. A remarkable Tlingit story (*Tlingit T.M.*, p. 227) tells how a resuscitated human being knows how potlatch has been carried out for him. The theme of the spirits who criticize the living for not having given potlatch is a common one. The Kwakiutl certainly had the same principles, e.g. the speech in *Ethn. Kwa.*, p. 227. Among the Tsimshian, the living represent the dead. Tate wrote to Boas: 'The offerings appear above all in the form of presents given at a festival' (*Tsim. Myth.* p. 452 (Historical legends), p. 287). See the collection of themes in Boas, ibid, p. 846 for the comparisons with the Haïda, Tlingit, and Tsimshian.

153   See below, n. 243 for further examples of the value of copper objects.
154   Krause, *Tlinkit Indianer*, p. 240, well describes the ways in which the Tlingit tribes approach one another.
155   Davy, 'Foi jurée', pp. 171 ff., 251 ff. The Tsimshian form does not differ perceptibly from the Haïda form, although perhaps the clan is more prominent.
156   It is otiose to go over the exposition of Davy regarding the relationship between the potlatch and political status, particularly that of the son-in-law and the son. It is likewise unnecessary to comment upon the ceremonial value of the banquets and exchanges. For example, the exchange of boats between two spirits causes them from then on to have no more than 'a single heart', the one being the father-in-law and the other the son-in-law (*Sec. Soc.*, p. 387). The text in *Kwa.*, vol. 3, p. 274 adds: 'it was as if they had exchanged their name'. See also ibid, vol. 3, p. 23: in a myth of the Nimkish festival (another Kwakiutl tribe), the wedding feast has as its purpose the enthronement of the girl in the village 'where she is going to eat for the first time'.
157   The funeral potlatch has been documented and sufficiently studied among the Haïda and the Tlingit. Among the Tsimshian it seems more particularly attached to the end of the mourning period, to the erection of the totem pole, and the cremation: *Tsim. Myth.*, p. 534 ff. Boas does not indicate that there is any funeral potlatch among the

Kwakiutl, but a description of a potlatch of this kind is to be found in a myth: *Kwa.*, vol. 3, p. 407.

158 A potlatch to uphold one's right to a coat of arms (Swanton, *Haïda*, p. 107). See the story of Legick, *Tsim. Myth.*, p. 386. Legick is the title of the principal Tsimshian chief. See also ibid, p. 364, for the stories of chief Nesbalas, the other great title of the Tsmishian chief, and the way he mocks chief Haïmas. One of the most important titles of chiefs among the Kwakiutl (Lewikilaq) is that of Dabend (*Kwa.*, vol. 3, p. 19, line 22, cf. *dabendgal'ala*, *Ethn. Kwa.*, p. 1406, col. 1), who, before the potlatch, has a name that means 'incapable of holding to the end', and after the potlatch takes a name that means 'capable of holding to the end'.

159 A Kwakiutl chief says: 'This is my vanity; the names, the roots of my family, all my ancestors have been . . . ' (and here he gives his name, which is at the same time a title and a common name) 'donors of *maxwa*' (a great potlatch): *Ethn. Kwa.*, p. 887, line 54. Cf. p. 843, line 70.

160 See below n. 209 (in a speech): 'I am covered with property. I am rich in property. I am the counter of property': *Ethn. Kwa.*, p. 1280, line 18.

161 To buy a copper object is to put it 'under the name' of the buyer: Boas, *Sec. Soc.*, p. 345. Another metaphor is that the name of the donor of the potlatch 'takes on weight' through the potlatch that is given (*Sec. Soc.*, p. 349), and 'loses weight' through a potlatch that is accepted (*Sec. Soc.*, p. 345). There are other expressions of the same idea of the superiority of donor over recipient: the notion that the latter is in some way a slave as long as he has not been 'bought back'. ('The name is bad', then, say the Haïda: Swanton, *Haïda*, p. 70. Cf. below, n. 203.) The Tlingit say that 'one puts the gifts on the back of the people that receive them' (Swanton, *Tlingit*, p. 428). The Haïda have two very symptomatic expressions of this 'to cause (his needle) to move', 'to run quickly' (cf. the New Caledonian expression p. 21), which apparently means 'to fight an inferior': Swanton, *Haïda*, p. 162.

162 See the story of Haïmas and how he lost his liberty, his privileges, his masks and other goods, his attendant spirits, his family, and his property: *Tsim. Myth.*, pp. 361, 362.

163 *Eth. Kwa.*, p. 805. Hunt, the Kwakiutl author of Boas, writes to him: I do not know why chief Maxuyalidze (meaning in fact, 'giver of the potlatch') never really gave a festival. That is all. Thus he was called Qelsem, namely, 'Rotten Face', ibid, lines 13–15.

164 The potlatch is in fact a dangerous thing, either because one is not forgiven [if one does not give one], or because one is its recipient. The persons who came to a mythical potlatch died of it (*Haïda T.M.*, Jesup Expedition, vol. 6, p. 626. Cf. p. 667, where there is the same Tsimshian myth). For comparisons, see Boas, *Indianische Sagen*, p. 356, no. 58. It is dangerous to partake of the substance of the one giving the potlatch, for example, to consume at a potlatch of spirits, in this world. This is a Kwakiutl legend (Awikenoq) (See *Ind. Sagen*, p. 239), as is also the beautiful myth of the crow who produces from its flesh various varieties of food: Çtatloq, *Ind. Sagen*, p. 76; Nootka, ibid, p. 106. Comparisons are given in Boas, *Tsim. Myth.*, pp. 694, 695.

165 The potlatch is in fact a game as well as a test. For example, the test consists in not having hiccups during the feast. 'Rather die than have hiccups', it is said. Boas, *Kwakiutl Indians*, Jesup Expedition, vol. 5, part 2, p. 428. There is a formula of challenge: 'Let us try to let our guests eat them [their plates] clean', *Ethn. Kwa.*, p. 991, line 43. Cf. p. 992. Concerning the uncertainty of the meaning between the words that signify the giving of food, the returning of food, and revenge, see Glossary (*Ethn. Kwa.*) under the heading *yenesa, yenka*: to give food, to recompense, to take one's revenge.

166 See above, n. 141 for the equivalence between the potlatch and war. The knife on the end of the stick is a symbol of the Kwakiutl potlatch: *Kwa.*, vol. 3, p. 483. Among the Tlingit, it is the raised spear (*Tlingit T.M.*, p. 117. See the rituals of the potlatch of compensation among the Tlingit. The war of the people of Kloo against the Tsimshian: *Tlingit T.M.*, pp. 432, 433, n. 34; dances after having made someone a slave; potlatch with no dance after having killed someone. The ritual of the giving of copper is dealt with elsewhere.

167 On ritual errors among the Kwakiutl, see Boas, *Sec. Soc.*, pp. 433, 507, etc. Expiation consists, in fact, of giving a potlatch or at least a gift.

In all societies this is a principle of law and ritual that is extremely important. A distribution of wealth plays the role of a fine, is a propitiation of the spirits and a re-establishment of community feeling with men. Fr Lambert, *Moeurs des sauvages néocalédoniens*, p. 66, has already noted among the Canucks the right of relations on the mother's side to claim indemnities when one of them loses blood in the family of his father. The self-same institution is to be found among the Tsimshian: Duncan, in Mayne, *Four Years*, p. 265. Cf.

p. 296 (potlatch in the case of the loss of blood of a son). The Maori institution of *muri* should probably be compared with this.

The potlatch for the purchasing back of prisoners should be interpreted in the same way. It is not only in order to get back the prisoner, but also to re-establish the 'name', that the family, which has let him become a slave, must give a potlatch. See the story of Dzebasa, *Tsim. Myth.*, p. 388. The same rule exists among the Tlingit: Krause, *Tlinkit Indianer*, p. 245; Porter, *Eleventh Census*, p. 54; Swanton, *Tlingit*, p. 449.

The potlatches for the expiation of ritual errors among the Kwakiutl are numerous. But one should note particularly the potlatch of expiation for parents of twins who are going off to work elsewhere, *Ethn. Kwa.*, p. 691. A potlatch is owed to a father-in-law to win back a wife who has left you, clearly through your own fault. See Vocabulary, ibid, p. 1423, col. 1, bottom. The principle can be used in an artificial way: when a chief seeks an opportunity for a potlatch, he sends his wife back to her father-in-law in order to have a pretext for fresh distributions of wealth: Boas, 'Fifth Report', p. 42.

168   A long list of these obligations to give festivals – after fishing expeditions, the gathering of food, hunting, the opening of boxes of preserves – is given in the first volume of *Ethn. Kwa.*, p. 757 ff. Cf. p. 607 ff. for etiquette, etc.

169   See above, n. 136.

170   See *Tsim. Myth.*, pp. 439, 512; cf. p. 534 for payment of services. A Kwakiutl example is payment for the one who counts the blankets, *Sec. Soc.*, pp. 614, 629 (Nimkish, the summer festival).

171   The Tsimshian have a remarkable institution that lays down the size of the shares at the potlatch of chiefs and the potlatch of vassals, and divides them out between one another. Although rivals confront one another within the different feudal classes, which are vertically divided into clans and phratries, there are, however, rights that are exerted from one class to another (Boas, *Tsim. Myth.*, p. 539).

172   Payments to parents. *Tsim. Myth.*, p. 534. Cf. Davy, 'Foi jurée' for opposing systems among the Tlingit and the Haïda, and the systems of distribution of potlatch by families, p. 196.

173   A Haïda myth given in Masset, *Haïda Texts*, Jesup, vol. 6, no. 43, tells how an old chief does not give enough potlatch. The others no longer invite him. He dies because of this. His nephews make a statue of him, give a festival, ten festivals in his name; he is then reborn. In another myth in Masset, ibid, p. 727, a spirit addresses a

chief and says to him, 'You have too much property, you must make
a potlatch (*wal* = distribution; cf. *walgal* = potlatch) of it.' He has a
house built and pays the builders. In another myth (ibid, p. 723, line
34) a chief says, 'I shall keep nothing for myself.' Cf. later, 'I shall
make a potlatch (*wal*) ten times.'

174   On the way that clans regularly confront one another (Kwakiutl), see
Boas, *Sec. Soc.*, p. 343; (Tsimshian), Boas, *Tsim. Myth.*, p. 497. In
areas where phratries exist this is taken for granted. See Swanton,
Haïda, p. 162; *Tlingit*, p. 424. This principle is remarkably well
expounded in the myth of the Crow (see *Tlingit T.M.*, p. 115 ff).

175   Naturally one refrains from inviting those who have been unworthy,
those who have given no festivals, those who do not possess the
'names' of festivals. See Hunt in *Ethn. Kwa.*; p. 707 for those that
have not returned the potlatch. Cf. ibid, Index, under the headings
Waya and *Wayapo Lela*, p. 1395; cf. p. 358, line 25.

176   Hence the constantly recurring story – equally common to both
European and Asiatic folklore – of the danger that exists in not inviting the orphan, the person abandoned by everybody, and the poor
person who turns up unexpectedly. E.g., *Indianische Sagen*, pp. 301,
303; see *Tsim. Myth.*, pp. 292, 295 for a beggar who is the totem, the
god of the totem. For a catalogue of the themes see Boas, *Tsim.
Myth.*, p. 784 ff.

177   The Tlingit have a remarkable expression: the guests are alleged to
'float', their boats 'wander over the sea', the totem pole that they
bring is being borne along: it is the potlatch, the invitation, that halts
them (*Tlingit Myth.*, p. 394, no. 22, p. 395, no. 24 (in speeches)). One
fairly common title of the Kwakiutl chief is 'the one towards whom
one paddles', he is 'the place to where one comes'. For example, see
Ethn. Kwa., p. 187 lines 10, 15.

178   The offence that consists in neglecting someone means that his
parents, out of solidarity, refrain also from coming to the potlatch. In
a Tsimshian myth, the spirits do not come so long as the Great Spirit
has not been invited, they all come when he is invited, *Tsim. Myth.*, p.
277. A story tells how when the great chief Nesbalas had not been
invited, the other Tsimshian chiefs did not come; they said: 'He is the
chief, we must not get on bad terms with him' (ibid, p. 357).

179   The offence has political consequences. For example, the potlatch of
the Tlingit with the Eastern Athapascans. Swanton, *Tlingit*, p. 435. Cf.
Tlingit T.M., p. 117.

180   *Tsim. Myth.*, pp. 170, 171.

181 Boas puts this piece of the text by Tate, his native reporter, as a note (ibid, p. 171, note [a]). On the contrary, the morality of the myth must be combined with the myth itself.

182 Cf. the details of the Tsimshian myth of Negunaks, ibid, p. 287 ff., and the notes on p. 846 for the equivalent of this theme.

183 E.g. The invitation to the blackcurrant feast, when the herald says: 'We invite you, you who have not come', *Ethn. Kwa.*, p. 752.

184 Boas, *Sec. Soc.*, p. 543.

185 Among the Tlingit the guests who have delayed two years before coming to the potlatch to which they were invited are called 'women', *Tlingit T.M.*, p. 119, n. [a].

186 Boas, *Sec. Soc.*, p. 345.

187 Kwakiutl. One is obliged to come to the feast of the seals, although the fat causes one to vomit: *Ethn. Kwa.*, p. 1046. Cf. p. 1048: 'try to eat everything'.

188 This is why sometimes one addresses oneself to one's guests in fear, for if they rejected the invitation, it is because they would show themselves to be superior. A Kwakiutl chief says to a Koskimo chief (tribe of the same nation): 'Do not refuse my kind offer, or I shall be ashamed, do not reject my heart . . . I am not one of those who make claims, or those who only give to those who will buy from them (= will give). So is it, my friends' (Boas, *Sec. Soc.*, p. 546).

189 Boas, *Sec. Soc.*, p. 355.

190 See *Ethn. Kwa.*, p. 774 ff. for another description given of the festival of the oils and the berries. It is by Hunt and seems a better one. It also appears that this ritual is employed in cases where one does not invite nor does one give. A festival ritual of the same kind, given to put a rival to scorn, includes singing to the accompaniment of the drum (ibid, p. 770; cf. p. 764), as among the Eskimos.

191 A Haïda formula: 'Do the same thing, give me good food' (in a myth), *Haïda Texts*, Jesup vol. 6., pp. 685, 686. (Kwakiutl), *Ethn. Kwa.*, p. 767, line 32, p. 738, line 32, p. 770, the story of PoLelasa (sic).

192 Songs denoting that one is not satisfied are very precise. Tlingit, *Tlingit T.M.*, p. 396, nos 26, 29.

193 Among the Tsimshian the chiefs have a rule of sending a messenger to examine the presents that those invited to the potlatch bring with them (*Tsim. Myth.*, p. 184; cf. pp. 430, 434). According to a capitular decree of 803, at Charlemagne's court there was an official entrusted with an inspection of this kind. Maunier drew my attention to this fact, which is mentioned by Demeunier.

194 See above, n. 161. Cf. the Latin expression *aere obaratus*, 'indebted'.

195 The myth of the Crow among the Tlingit tells how the latter is not at a festival because the others (the opposing phratry) have shown themselves to be noisy, and have crossed the dividing line that in the dance-house separates the two phratries. The Crow feared that they were invincible (*Tlingit T.M.*, p. 118).

196 The inequality that is the consequence of the fact of accepting is well brought out in Kwakiutl speeches. *Sec. Soc.*, pp. 355, 667, line 17. Cf. p. 669, line 9.

197 E.g. the Tlingit. Swanton, *Tlingit*, pp. 440, 441.

198 Among the Tlingit a ritual allows one to contrive to be paid more and on the other hand allows the host to force a guest to accept a present. The dissatisfied guest makes a move to leave. The donor offers him double the amount, mentioning the name of a dead relative (Swanton, *Tlingit Indians*, p. 442). It is probable that this ritual corresponds to the right that the two contracting parties have to represent the spirits of their ancestors.

199 See speech, *Ethn. Kwa.*, p. 1281: 'the chiefs of the tribes never give in return . . . they disgrace themselves, and you raise yourself up as a great chief, among those who have disgraced themselves'.

200 See the speech (a historical story) at the potlatch of the great chief Legek (the title of the prince of the Tsimshian), *Tsim. Myth.*, p. 386. They say to the Haïda, 'You will be the least among the chiefs because you are not capable of casting away into the sea the copper objects, as the great chief has done.'

201 The ideal would be to give a potlatch that is not returned. See in a speech: 'You wish to give what will not be returned' (*Ethn. Kwa.*, p. 1282, line 63). The individual who has given a potlatch is compared to a tree or a mountain: 'I am the great chief of the great tree, you are beneath me . . . my fence . . . I give you property' (ibid, p. 1290, verse 1). 'Lift the pole of the potlatch, what is unattackable is the only thick-trunked tree, the only thick root' (ibid, verse 2). The Haïda express this in the metaphor of the spear. The people that accept 'live from his (the chief's) spear' (*Haïda Texts* (Masset) p. 486). It is moreover one type of myth.

202 See the story of an insult for a potlatch that was badly reciprocated (*Tsim. Myth.*, p. 314). The Tsimshian always remember the two copper objects that are due to them from the Wutsenaluk (ibid, p. 364).

203 The 'name' remains 'broken', so long as one has not broken a copper object equal in value to that of the challenge (Boas, *Sec. Soc.*, p. 543).

204 When an individual who has been discredited in this way borrows
something in order 'to make a distribution' or 'to make an obligatory
redistribution', he 'commits his name', and the synonymous expression is 'he sells a slave' (Boas, *Sec. Soc.*, p. 341). Cf. *Ethn. Kwa.*,
pp. 1424, 1451 under the heading *kelgelgend*. Cf. p. 1420.

205 His intended bride may not yet be born, but the contract already
binds the young man (Swanton, *Haïda*, p. 50).

206 See above, n. 132. In particular, the peace rituals among the Haïda,
the Tsimshian, and the Tlingit consist of total instant services and
total counter-services: in fact, exchanges of pledges (emblazoned
copper objects) and hostages, both slaves and women. For example,
in the war of the Tsimshian against the Haïda (see *Haïda T.M.*, p.
395): 'As they had marriage of women with their opponents on both
sides and because they feared that anger could be stirred up again,
there was peace.' In a war of the Haïda against the Tlingit, for a
potlatch of compensation, see ibid, p. 396.

207 See above, n. 170, and especially Boas, *Tsim. Myth.*, pp. 511, 512.

208 (Kwakiutl): a distribution of property, one piece after another, from
both sides (Boas, *Sec. Soc.*, p. 418); repayment the following year of
fines paid for errors in ritual: ibid, p. 596; repayment with interest of
the bride-price, ibid, pp. 365, 366, 423, line 1, 518–50, 563.

209 On the word potlatch see above, Intro., n. 13, p. 86. Moreover, it
seems that neither the idea nor the nomenclature behind the use of
this term have in the languages of the Northwest the kind of precise-
ness that is afforded them in the Anglo-Indian 'pidgin' that has
Chinook as its basis.

In any case Tsimshian distinguishes between the *yaok*, a great
intertribal potlatch (Boas, [Tate], *Tsim. Myth.*, p. 537; cf. p. 511, cf. p.
968, incorrectly translated by potlatch) and the others. The Haïda
distinguish between the *walgat* and the *sika* (Swanton, *Haïda*, pp. 35,
178, 179; p. 68 (Masset's text), the funeral potlatch, and a potlatch
for other reasons.

In Kwakiutl the word common to both Kwakiutl and Chinook, *poLa*
(sic) ('to eat one's fill') (*Kwa.*, vol. 3, p. 211, line 13, *PoL*, 'sated', ibid,
vol. 3, p. 25, line 7) seems not to designate the potlatch, but the feast
or the effect of the feast. The word *poLas* designates the giver of the
feast (*Kwa. T.*, 2nd Series, Jesup, vol. 10, p. 79, line 14; p. 43, line 2)
and also designates the place where one is satiated (legend of the
title of one of the Dzawadaenoxu chiefs). Cf. *Ethn. Kwa.*, p. 770, line
30. The most general name in Kwakiutl, is *p!Es*, 'to flatten' (the name

of one's rival) (Index, *Ethn. Kwa.*, under this heading), or it can mean 'baskets being emptied' (*Kwa.*, vol. 3, p 93, line 1; p. 451, line 4). The great tribal and intertribal potlatches seem to have their own special name, *maxwa* (*Kwa.*, vol. 3, p. 451, line 15). From the root of *ma* Boas derives two other somewhat unlikely words: the one is *mawil*, the initiation room, and the other the name of the killer whale (*Ethn. Kwa.*, Index). In fact among the Kwakiutl one finds a host of technical terms to designate all kinds of potlatch and also each of various kinds of payments and repayments, or rather, gifts and reciprocal gifts: for weddings, for payments to the Shaman, for advances, for a backlog of interest – in short, for all kinds of acts of distribution and redistribution. An example is: *men(a)*, 'pick up', (*Ethn. Kwa.*, p. 218): 'A small potlatch in which the clothing of a young girl is thrown to the people and picked up by them; *payol*, 'to give a copper object'; another term for giving a boat (*Ethn. Kwa.*, p. 1448). The terms are numerous, varying, and concrete, and overlap with one another, as in all kinds of archaic nomenclature.

210   See Barbeau (1911) 'Le potlatch', *Bull. Soc. Géog. Québec*, vol. 3, p. 278, n. 3 for this meaning and the references indicated.

211   Perhaps also, for sale.

212   The distinction between property and provisions is very clear in Tsimshian. In *Tsim. Myth.*, p. 435, Boas says, doubtless following Tate, his correspondent: 'The possession of what is called rich food (cf. ibid, p. 406) was essential to maintain the family nobility. But provisions were not counted as constituting wealth. Wealth was obtained through sale (we would really say, gifts exchanged) of provisions or other kinds of goods, which, having been accumulated, were given out at the potlatch (see above, n. 111).

   The Kwakiutl also distinguish between mere provisions and property wealth. These last two words are equivalent. The latter seems to have two names (see *Ethn. Kwa.*, p. 1454). The first is *yag* [? text unreadable: trans.] or *yaq* – Boas' philology varies. (See index, p. 393 under this entry.) The word has two derivatives: *yeqala*, property, and *yaxulu*, talisman goods and paraphernalia. Cf. the derived words from *yä*, ibid, p. 1406. The other word is *dadekas*, cf. Index to *Kwa. T.*, vol. 3, p. 519. Cf. ibid, p. 473, line 31; in Newettee dialect, *daoma*, *dedemala* (see Index to *Ethn. Kwa.*, under this heading). The root of this word is *dâ*. The latter has meanings curiously analogous to those of the identical root *dâ* in Indo-European: to receive, to take, to carry in one's hand, to manipulate, etc. Even the derivatives are significant. One means 'to take a piece of the enemy's clothing in order to cast a spell upon him', another 'to put in hand', 'to put in the home' (for a comparison of the meanings of *manus* and *familia* see elsewhere (concerning blankets given before the purchase of copper objects, to be returned with interest). Another word means 'to put a quantity of blankets on top of the pile of one's opponent, and to accept them', by performing this gesture. A derivative from the same root is even more curious: *dadeka*, 'to be jealous of one another' (see *Kwa. T.*, p. 133, line 22). Evidently the original meaning must be: the thing that one takes and that causes jealousy; cf. *dadego*, 'to fight', doubtless to fight through property.

Other words still have the same meaning, but are more precise. For example, *mamekas*, 'property in the house' (*Kwa. T.*, vol. 3, p. 169, line 20).

213   See various speeches at which the handing over is made (Boas and Hunt, *Ethn. Kwa.*, p. 707 ff.).

There is almost nothing that is morally and materially valuable (intentionally we do not employ the word 'useful') that is not an object of beliefs of this kind. Firstly, in fact, moral things are goods, property, the object of gifts and exchanges. For example, as in more primitive civilizations – those of Australia, for instance – one leaves the *corroborree*, the representation that has been learnt from it, with the tribe to whom it has been handed over just as, among the Tlingit after the potlatch, one 'leaves' a dance with the people who gave it you (Swanton, *Tlingit Indians*, p. 442). The essential property among the Tlingit, the one that is the most inviolable and excites the jealousy of people, is the name and the totem emblem, ibid, p. 416. Moreover, it is this that makes one happy and rich.

Totem emblems, festivals and potlatch, the names won during these potlatches, the presents that others will have to reciprocate, and that are linked to the potlatches that have been given – all this follows. A Kwakiutl example is given in a speech: 'And now my festival goes to him' (pointing to the son-in-law (*Sec. Soc.*, p. 356)). It is the 'seats' and also the 'spirits' of secret societies that are given and returned in this way (see a speech on the ranks of property, and the property of ranks), *Ethn. Kwa.*, p. 472. Cf. ibid, p. 708, another speech: 'This is your winter song, your winter dance, everybody will take property on it, on the winter blanket; this is your song, this is your dance.' A single word in Kwakiutl designates the talismans of the noble family and its privileges: the word *kleso*, 'emblem, privilege'. For example, *Kwa. T.*, vol. 3, p. 122, line 32.

Among the Tsimshian the masks and emblazoned hats for dancing and parading are termed 'a certain quantity of property' according to the quantity given at the potlatch (according to the presents made by the maternal aunts of the chief to 'the women of the tribe'), Tate in Boas, *Tsim. Myth.*, p. 541.

Conversely, for example, among the Kwakiutl it is in a moral light that things are viewed, especially the two precious things, the essential talismans, the 'giver of death' (*halaya*) and the 'water of life' (evidently a single quartz crystal), and the blankets, etc. that we have already mentioned. In a curious Kwakiutl saying, all these paraphernalia are identified with the grandfather, as is natural, since they are only lent to the son-in-law in order to be given back to the grandson (Boas, *Sec. Soc.*, p. 507).

214 The myth of Djîlaqons is found in Swanton, *Haïda*, pp. 92, 95, 171. Masson's version is found in *Haïda T.*, Jesup, vol. 6, pp. 94, 98; that of Skidegate in *Haïda T.M.*, p. 458. The name figures among a certain number of Haïda family names belonging to the eagle phratry (see Swanton, *Haïda*, pp. 282, 283, 292, 293). In Masset the name of the goddess of fortune is Skîl: *Haïda T.*, Jesup, vol. 6, p. 665, line 28; p. 306. Cf. Index, p. 805; cf. the bird Skîl, Skirl (Swanton, *Haïda*, p. 120). *Skîltagos* means 'copper property', and the fabulous story of how the 'copper objects' were found is linked to that name (cf. p. 146, Fig. 4). A sculpted pole represents Djîlqada, its copper object, its pole, and its emblems (Swanton, *Haïda*, p. 125; cf. Plate 3, Fig. 3). See the descriptions by Newcombe, ibid, p. 46. Cf. the figurative representation, ibid, Fig. 4. Its fetish must be crammed with stolen things, and itself stolen.

Its exact title is 'property making a noise' (ibid, p. 92). It has four additional names (ibid, p. 95). It has a son who bears the title 'stone ribs' (in reality, of copper (ibid, pp. 110, 112)). He who meets it, or its son or daughter, is lucky in gambling. It has a magic plant. One becomes rich if one eats of it. One also becomes rich if one touches a piece of its blanket, and if one finds mussel shells that it has laid out in order (ibid, pp. 29, 109).

One of its names is 'property keeping in the house'. A large number of people hold titles made up of Skîl: 'He who awaits Skîl', 'the way towards Skîl'. See in the genealogical lists of the Haïda, E. 13, 14; and the phratry of the Crow, R. 12, R. 15, R. 16.

It would seem to be the opposite of the 'plague woman' (cf. *Haïda T.M.*, p. 299).

215 On the Haïda *djîl* and the Tlingit *näq* see above, n. 138.

216 The myth is to be found in complete form among the Tlingit (*Tlingit T.M.*, pp. 173, 292, 368). Cf. Swanton, *Tlingit*, p. 460. At Sitka the name of Skîl is undoubtedly Lennaxxidek. It is a woman who has a child. One hears the noise of this child being suckled. One runs after the child. If one is scratched by him and one has scars, the scabs of these make others happy.

217 The Tsimshian myth is incomplete (*Tsim. Myth.*, pp. 154, 197). Compare the notes of Boas, ibid, pp. 746, 760. Boas has not made the identification, but it is clear. The Tsimshian goddess wears 'a garment of wealth'.

218 It is possible that the myth of the Qominoqa, of the 'rich woman', has the same origin. She seems to be the object of a cult reserved for certain clans among the Kwakiutl, e.g. *Ethn. Kwa.*, p. 862. A hero of the Qoexsotenoq bears the title of 'body of stone' and becomes 'property on the body' (*Kwa. T.*, vol. 3, p. 187). Cf. p. 247.

219 See, for example, the myth of the clan of the killer whales (Boas, *Handbook of American Languages*, vol. 1, pp. 554–9). The hero author of the clan is himself a member of the killer-whale clan: 'I am seeking to find a *logwa* (a talisman, cf. p. 554, line 49) of you', he says to a spirit of a killer whale he meets, who has human form, p. 557, line 122. The latter recognizes him as being of his clan; he gives him the harpoon with the copper hook that kills whales (forgotten in the text, p. 557): the clan are killer whales. The spirit also gives him his potlatch name. He will be called 'place to eat one's fill', 'feeling himself sated'. His house will be the 'house of the killer whale', with 'one painted on its front'. 'And killer whale shall be your dish in the house (it will take the form of killer whale), and also the *halayu* (the meter-out of death), and "water of life" and a knife with quartz teeth shall be your cutting-knife (of whales)' (p. 559).

220 A miraculous box containing a whale, one that has given its name to a hero, bore the title of 'riches coming to the shore' (Boas, *Sec. Soc.*, p. 374). Cf. 'property is drifting towards me', ibid, pp. 247, 414. The property 'makes a noise' (see above, n. 214). The title of one of the principal chiefs in Masset is 'He whose property makes a noise' (*Haïda Texts*, Jesup, vol. 6, p. 684). The property is alive (Kwakiutl): 'May our property remain alive under its efforts, and our copper object remain unbroken', sing the Maamtagila, *Ethn. Kwa.*, p. 1285, line 1.

221 The paraphernalia of the family, those that circulate among men, their daughters, and their sons-in-law, and come back to the sons when they have been recently initiated or get married, are normally contained in a box or case, decorated and emblazoned. The assembly and fashioning of the case, as well as its use, are entirely characteristic of this civilization of the American Northwest (from the Yurok of California to the Behring Straits). This box is generally adorned with the faces and eyes either of totems or of spirits, whose attributes it enshrines: these are ornamented blankets, the talismans of 'life' and 'death', the masks, the mask-hats, the hats and the crowns, and the bow. Mythology often mixes up the spirit with this box and its contents, e.g. *Tlingit T.M.*, p. 173: the *gonaqadet*, which is identical with the box, the copper object, the hat, and the rattle with a bell.

222 It is the act of its transfer, the gift made to him, that originally, as at each fresh initiation or marriage, transforms the recipient into a 'supernatural' person, into an initiate, a Shaman, a magician, a noble, the one who possesses dances and seats in a brotherhood (see the speeches in the histories of the Kwakiutl families, *Ethn. Kwa.*, pp. 965, 966; cf. p. 1012).

223 The miraculous box is still mysterious, and is preserved in the secret places of the house. There can be boxes within boxes, each placed inside the other in great number (Haïda), Masset, *Haïda Texts*, Jesup, vol. 6, p. 395. It contains spirits, for example, the 'mouse woman' (Haïda) (*Haïda T.M.*, p. 340). Another example is: the crow that pecks the eyes out of the faithless person who has it in his possession. See the catalogue of examples of this theme in Boas, *Tsim. Myth.*, pp. 854, 851. The myth of the sun shut up in the box, which is floating, is one of the most widespread myths (list given in Boas, *Tsim. Myth.*, pp. 549, 641). The extension of these myths into the ancient world is well known.

One of the commonest episodes in the stories of the hero is that of the very small box, fairly heavy for him, too heavy for anyone else to carry, in which there is a whale (Boas, *Sec. Soc.*, p. 374; *Kwa. T.*, 2nd series, Jesup, vol. 10, p. 171), whose food is inexhaustible (ibid, p. 223). This box is alive and floats of its own accord (*Sec. Soc.*, p. 374). The box of Katlian brings riches (Swanton, *Tlingit Indians*, p. 448; cf. p. 446). The flowers, the 'dung of the sun', 'egg of kindling wood', 'who make one rich' – in other words, the talismans they contain, the wealth itself, must be fed.

One box contains the spirit 'too strong to be taken over', whose mask kills the wearer (*Tlingit T.M.*, p. 341).

The names of these boxes are often symptomatic of their use at the potlatch. A large Haïda box for fat is called the mother (Masset), *Haïda Texts*, Jesup, vol. 6, p. 758. The 'box with the red bottom' (the sun) 'sheds water' into the 'sea of the tribes' (the water is represented by the blankets that the chief distributes) (Boas, *Sec. Soc.*, p. 551 and n. 1; p. 564).

The mythology of the miraculous box is also characteristic of the societies of the North-Asian Pacific. A fine example of a comparable myth is to be found in Pilsudski (1913) *Material for the Study of the Atnu Languages*, Cracow, p. 124 and p. 125. This box is given by a bear, and the hero must observe taboos. It is full of articles of gold and silver, and talismans that give wealth. The technique of the box is, moreover, the same throughout the North Pacific.

224 The 'things of the family are individually named' (Haïda), Swanton, *Haïda*, p. 117. The following have names: houses, doors, dishes, carved spoons, boats, salmon traps. Cf. the expression 'continuous chain of properties' (Swanton, *Haïda*, p. 15).

We possess the list of things that are given names by the Kwakiutl according to clans, as well as the titles (which vary) given to nobles, both men and women, and their privileges, such as dances and potlatches, which are likewise properties. Things that we would call furniture, and that are given names and personified under the same conditions are: dishes, the house, the dog, and the boat (see *Ethn. Kwa.*, p. 793 ff.). In this list Hunt has omitted to mention the names of copper objects, large abalone shells, and doors. The spoons threaded to a rope tethered to a kind of boat with carved figures are called 'the spoon anchor line' (see Boas, *Sec. Soc.*, p. 442, in a ritual for paying off a marriage debt). Among the Tsimshian the following are given names: boats, copper objects, spoons, stone pots, stone knives, the dishes of female chiefs (Boas, *Tsim. Myth.*, p. 506). The slaves and the dogs are always things of value and are creatures adopted by families.

225 The only domestic animal of these tribes is the dog. It bears a different name according to the clan (and is probably in the chief's family). It cannot be sold. 'They are men, like us', say the Kwakiutl (*Ethn. Kwa.*, p. 1260). 'They protect the family' against witchcraft and the attacks of enemies. A myth tells how a Koskimo chief and Waned, his dog, could change into each other and bore the same name

(ibid., p. 835, cf. Ch. 4, n. 4, p. 154 and (Celebes); cf. the fantastic myth of the four dogs of Lowiqilaku, *Kwa.*, vol. 3, pp. 18, 20).

226 *Abalone* is the Chinook patois word to designate the large *haliotis* shells that serve as ornaments – nose rings (Boas, *Kwa. Indians*, Jesup, vol. 5, part 1, p. 484), and earrings (Tlingit and Haïda (see Swanton, *Haïda*, p. 146)). They are also attached to emblazoned blankets, to belts and hats, e.g Kwakiutl (*Kwa.*, p. 1069). Among the Awikenoq and the Lasiqoala (tribes of the Kwakiutl group) the abalone shells are arranged round a shield, one that is strangely European in shape (Boas, 'Fifth Report', p. 43). This kind of shield seems to be the primitive form, or the equivalent of the copper shields, the shape of which is also strangely like those of the Middle Ages.

It seems that the abalone shells must have once had value as money, of the same kind as that of the copper objects today. A Ctatlolq myth (South Salish), associates the two characters, K'okois, 'copper' and Teadjas, 'abalone'. Their son and daughter marry and the grandson takes the 'metal box' of the bear, seizing possession of its mask and its potlatch (*Indianische Sagen*, p. 84). An Awikenoq myth links the names of shells, like those of the different forms of copper, to 'daughters of the moon' (ibid, pp. 218, 219).

Among the Haïda these shells each have their own name, at least when they have great value and are well known, just as do they in Melanesia (cf. Swanton, *Haïda*, p. 146). Elsewhere they serve to give names to individuals or spirits. For example, among the Tsimshian (see index of proper names, Boas, *Tsim. Myth.*, p. 960). Cf. among the Kwakiutl, the 'abalone names', by clans (*Ethn. Kwa.*, pp. 1261–75), for the Awikenoq, Naqoatok, and Gwasela tribes. Here the custom is certainly international. The box of abalone of the Bella Kula (a box ornamented with pearls), is itself mentioned and exactly described in the Awikenoq myth. Moreover, it contains the abalone blanket, and both have the brilliance of the sun. The name of the chief whose myth includes the story is Legek (Boas, *Ind. Sag.*, p. 218 ff.). This name is that of the principal Tsimshian chief. We realize that the myth has travelled with the thing. In a Haïda myth in Masset, that of 'the Crow the Creator' itself, the sun that he gives to his wife is an abalone shell (Swanton, *Haïda Texts*, Jesup, vol. 6, pp. 227, 313). For the names of mythical heroes bearing abalone titles, see examples in *Kwa. T.*, vol. 3, pp. 50, 222.

Among the Tlingit these shells were associated with sharks' teeth (*Tlingit T.M.*, p. 129). Compare the use of whales' teeth in Melanesia.

In addition, all these tribes have the cult of necklaces made of *dentalia* (small shells) (see in particular, Krause, *Tlinkit Indianer*, p. 186). All in all, we find here precisely the same forms of money, with the same beliefs and serving the same usage as in Melanesia, and in the Pacific in general.

These various shells were, moreover, the object of trading that was also carried on by the Russians during their occupation of Alaska. This trade went in both directions: between the gulf of California and the Behring Straits (Swanton, *Haïda Texts*, Jesup, vol. 6, p. 313).

227 The blankets are ornamented just like the boxes. Often they are even copied from the designs on the boxes (see Figure in Krause. *Tlinkit Indianer*, p. 200). There is always something witty about them (cf. the expressions (Haïda), 'belts of wit', torn blankets, Swanton *Haïda*, Jesup, vol. 5, part 1, p. 165; cf. p. 174). A certain number of mythical coats are 'coats of the world': (Lilloët, myth of Qäls, Boas, *Ind. Sagen*, pp. 19, 20); (Bella Kula) 'coats of the sun' (*Ind. Sagen*, p. 260); a coat with fish ornaments (Heiltsuq) (*Ind. Sagen*, p. 248). For a comparison of samples of this theme, see Boas, ibid, p. 359, no. 113.

Cf. the mat that talks (*Haïda Texts*, Masset, Jesup Expedition, vol. 6, pp. 430, 432). The cult of blankets and matting, of skins converted into blankets, makes it seem that it should be compared with the cult of the emblazoned mats in Polynesia.

228 Among the Tlingit it is accepted that everything talks in the house: the spirits talk to the posts and beams of the house; they speak from the posts and beams; the posts and beams speak; and dialogues are exchanged in this way between the totem animals, the spirits, and the men and the things in the house. This is a regular principle in Tlingit religion, e.g. Swanton, *Tlingit*, pp. 458, 459. Among the Kwakiutl the house listens and talks *Ethn. Kwa.*, p. 1279, line 14).

229 The house is conceived of as a kind of movable good. (We know that in Germanic law it remained so for a long time.) It is transported, and it transports itself (see the very numerous myths of the 'magic house', built in the twinkling of an eye, especially given by the grandfather (listed by Boas, *Tsim. Myth.*, p. 852, p. 853).) For Kwakiutl examples, see Boas, *Sec. Soc.*, p. 376 and the figures and plates on pp. 376, 380.

230 The following are likewise precious things, both magic and religious: [1] eagle feathers, often identified with rain, food, quartz, and 'good medicine', for example, *Tlingit T.M.*, pp. 128, 383; Haïda. (Masset), *Haïda Texts*, Jesup, vol. 6, p. 292; [2] sticks and combs (*Tlingit T.M.*,

p. 385; Haïda, Swanton, *Haïda*, p. 38; Boas, *Kwakiutl Indians*, Jesup, vol. 5, part 2, p. 455) [3] bracelets, e.g. a tribe of the Lower Fraser (Boas, *Indianische Sagen*, p. 36); (Kwakiutl), Boas, *Kwa. Ind.*, Jesup, vol. 5, part 2, p. 454).

231 All these objects, including the spoons, the dishes, and the copper objects, are included in Kwakiutl under the generic term of *logwa*, whose precise meaning is 'talisman, supernatural thing' (see the observations we have made on this word in our study, *Origines de la notion de monnaie*, and in our Preface to Hubert and Mauss, *Mélanges d'histoire des religions*). The notion of *logwa* is exactly that of *mana*. But in the event, and as regards the subject we are studying, it is the 'virtue' in wealth and food that produces wealth and food. One speech talks of a talisman, of the *logwa* that is 'the great increaser in the past of property' (*Ethn. Kwa.*, p. 1280, line 18). A myth tells how a *logwa* was 'happy to acquire property', how four *logwa* (belts, etc.) amassed it. One of them was called 'the thing that makes property accumulate' *Kwa. T.*, vol. 3, p. 108). In reality it is wealth that makes wealth. A Haïda saying even talks of 'property that makes one rich', concerning the abalone shells that a girl at puberty wears (Swanton, *Haïda*, p. 48).

232 A mask is called 'obtaining food' (cf. 'and you will be rich in food' (Nimkish myth), *Kwa. T.*, vol. 3, p. 36, line 8). One of the most important Kwakiutl nobles bears the title of 'the Inviter', that of the 'giver of food', that of the 'giver of the eagle's down' (cf. Boas, *Sec. Soc.*, p. 415).

The baskets and the ornamented boxes (for example, those used for the gathering of berries) are also magical, for example, Haïda myth (Masset), *Haïda T.*, Jesup, vol. 6, p. 404; the very important myth of Qäls mixes up the pike, the salmon and the thunderbird, and a basket that a drop of saliva from this bird fills with barries (tribe of the Lower Fraser River), *Ind. Sag.*, p. 34; the equivalent Awikenoq myth ('Fifth Report', p. 28) refers to a basket that bears the name 'never empty'.

233 The dishes are each named according to what the carving on them represents. Among the Kwakiutl they represent the 'animal chiefs' (cf. p. 44). One bears the title of 'dish that keeps full' (Boas, *Kwakiutl Tales* (Columbia University), p. 264, line 11). Those of a certain clan are *logwa*: they have spoken to an ancestor, 'the Inviter' (see note 232), and have told him to take them (*Ethn. Kwa.*, p. 809; cf. the myth of Kaniqilaku, *Ind. Sag.*, p. 198; cf. *Kwa. T.*, 2nd Series, Jesup, vol. 10,

p. 205): it tells how the transformer has given his father-in-law (who was tormenting him) some magic berries from his basket to eat. These transform themselves into a bramble bush and sprout out all over his body.

234   See above, n. 224.

235   Ibid.

236   The expression is borrowed from the German language, *Renommiergeld*, and was used by Krickeberg. It describes very precisely the use of these crown-shaped shields, sheets of metal that are at the same time coins and, especially, objects for display that the chiefs carry at the potlatch, or those in whose honour the potlatch is given.

237   Although much discussed, the copper industry in the American Northwest is still not well-enough known. Rivet (1923) in his remarkable study, 'L'Orfèvrerie précolombienne', *Journal des Américanistes*, expressly did not deal with it. In any case it seems certain that this art predates the arrival of Europeans. The Tlingit and the Tsimshian, the Northern tribes, searched for, mined, or received indigenous copper from the Copper River (cf. the older authors and Krause, *Tlinkit Indianer*, p. 186). All these tribes speak of the 'great mountain of copper' (Tlingit) *Tlingit T.M.*, p. 160; (Haïda), Swanton, *Haïda*, Jesup, vol. 5, p. 130; (Tsimshian), *Tsim. Myth.*, p. 299).

238   We take the opportunity to correct a mistake that we made in our *Note sur l'origine de la notion de monnaie*. We confused the word *laqa*, *laqwa* (Boas uses both spellings), with *logwa*. Our excuse is that at that time Boas often wrote the two words in the same way. However, since then, it has become clear that the one means 'red', 'copper', and the other means solely, 'a supernatural thing, a thing to be prized, a talisman'. All copper objects are, however, *logwa*, which means that our analysis remains valid. But in that case the word is a kind of adjective and synonym, e.g. in *Kwa. T.*, vol. 3, p. 108, two titles of *logwa* are given to copper objects: 'that which is happy to acquire property', and 'that which causes property to accumulate'. But all *logwa* are not copper objects.

239   Copper is regarded as being a living thing: its mine, its mountain, are magical, full of 'plants of wealth' (Masset, *Haïda Texts*, Jesup, vol. 6, pp. 681, 692; cf. Swanton, *Haïda*, p. 146, another myth). It is true that it has a smell (*Kwa.*, vol. 3, p. 64, line 8). The privilege of working the copper is the subject of an important cycle of legends among the Tsimshian: the myth of Tsauda and Gao (*Tsim. Myth.*, p. 306 ff.). For the list of equivalent themes, see Boas, *Tsim. Myth.*, p. 856. Copper

seems to have been personalized among the Bellakula, (*Ind. Sagen*, p. 306 ff.; cf. Boas, *Mythology of the Bella Coola Indians*, Jesup Exp., vol. 1, part 2, p. 71), where the copper myth is associated with the myth of the abalone shells. the Tsimshian myth of Tsauda is linked to the myth of the salmon, which we will discuss later.

240   Because it is red, copper is identified with [1] the sun, e.g. *Tlingit T.M.*, nos 39, 81; [2] a 'fire falling from heaven' (name of a type of copper), Boas, *Tsimshian Texts and Myths*, p. 467; [3] salmon, in any case. This identification is particularly clear in the case of the cult of twins among the Kwakiutl, a people of salmon and copper, *Ethn. Kwa.*, p. 685 ff. The mythical sequence seems to be as follows: springtime, the arrival of the salmon, a new sun, a red colour, copper. The identity between copper and salmon is more marked among the northern nations (see list of equivalent cycles, Boas, *Tsim. Myth.*, p. 856). For example, see the Haïda myth in Masset, *Haïda T.*, Jesup, vol. 6, pp. 689; 691, line 6 ff., n. 1 (cf. p. 692, myth no. 73). One finds here an exact equivalent of the legend of Polycrates' ring: that of a salmon that has swallowed a copper object (Skidegate, H., *T.M.*, p. 82). The Tlingit possess – as did the Haïda subsequently – the myth of the creature whose name is Mouldy-end (the name of a salmon); see the myth of Sitka: chains of copper and salmon (*Tlingit T.M.*, p. 307). A salmon in a box becomes a man – another version of Wrangell, ibid, no. 5. For equivalents see Boas, *Tsim. Myth.*, p. 857. A Tsimshian copper object bears the title of 'copper going back up a river', a clear allusion to the salmon (Boas, *Tsim. Myth.*, p. 857).

It would be rewarding to investigate what makes this cult of copper analogous to that of quartz, e.g. the myth of the quartz mountain, *Kwa. T.* 2nd Series, Jesup, vol. 10, p. 111.

In the same way, the cult of jade, at least among the Tlingit, must be compared with that of copper: a jade salmon speaks (*Tlingit T.M.*, p. 5). A jade stone speaks and gives names (Sitka, *Tlingit T.M.*, p. 416). Finally, we should recall the cult of shells and its associations with that of copper.

241   We have seen that the family of the Tsauda, among the Tsimshian, seem to be smelters of copper, or keepers of its secrets. It appears that the myth (Kwakiutl) of the princely family of Dzawadaenoqu is one of the same kind. It associates together: Laqwagila, 'the worker of copper', with Qomqomgila, 'the Rich One', and Qomoqoa, 'the Rich' (female) who makes copper objects (*Kwa.*, vol. 3, p. 50); it links the whole with a white bird (the sun), the son of the thunderbird that

smells of copper, and changes into a woman who gives birth to twins who smell of copper (*Kwa.*, vol. 3, pp. 61–7).

The Awikenoq myth relating to ancestors and nobles bearing the same title of 'worker of copper' is much less interesting.

242 Each copper object has its name. 'The big copper objects that have names', say Kwakiutl speeches, (Boas, *Sec. Soc.*, pp. 348, 349, 350). There is a list of the names of copper objects, but unfortunately no indication as to which clan is their permanent owner (ibid, p. 344). We are fairly well informed about the names of the large copper objects of the Kwakiutl. They show the cults and beliefs attached to them. One bears the title of 'Moon' (the Nisqa tribe) (*Ethn. Kwa.*, p. 856). Others bear the name of the spirit they embody and who gave them, e.g. the Dzonoqoa (*Ethn. Kwa.*, p. 1421). They represent its face. Others bear the name of founding spirits of totems: a copper object is called 'Beaver face' (*Ethn. Kwa.*, p. 1427); another, 'Sea lion' (ibid, p. 894). Other names merely allude to the shape: 'Copper of T-shape', or 'Long upper quarter' (ibid, p. 862). Others are simply called 'Big copper' (ibid, p. 1289), 'Resounding copper' (ibid, p. 962) (also the name of a chief). Other names refer to the potlatch that they embody, and whose worth is epitomized in them. The name of Maxtoselem, the copper object, is 'That of whom the others are ashamed' (cf. *Kwa.*, vol. 3, p. 452, n. 1: 'They are ashamed of their debts' (*gagim*)). Another name is 'Cause quarrel' (*Ethn. Kwa.*, pp. 893, 1026).

On Tlingit names of copper objects see Swanton, *Tlingit*, pp. 421, 405. Most of these names are totemic. For the names of Haïda and Tsimshian copper objects, we know only those that bear the same name as the chiefs, their owners.

243 The value of the copper objects among the Tlingit varies according to their height and is counted in numbers of slaves (*Tlingit T.M.*, pp. 131, 337, 260 (Sitka and Skidegate, etc. Tsimshian); Tate, in Boas, *Tsim. Myth.*, p. 540; cf. ibid, p. 436). For the equivalent principle: (Haïda), see Swanton, *Haïda*, p. 146.

Boas has studied closely the way in which each copper object increases in value with the series of potlatches: the value of the Lesaxalayo copper objects in about 1906–10 was: 9,000 woollen blankets, worth four dollars each, 50 boats, 6,000 buttoned blankets, 260 silver bracelets, 60 gold bracelets, 70 gold earrings, 40 sewing machines, 25 gramophones, 50 masks. The herald proclaims, 'For prince Laqwagila, I shall give away all these poor things.' (*Ethn.*

Kwa., p. 1352). Cf. ibid, line 28, in which the copper object is compared to the 'carcass of a whale'.

244   On the principle of destruction, see elsewhere. However, the destruction of copper objects seems to be of a particular character. Among the Kwakiutl it is done piece by piece, a new quarter being smashed at each potlatch. And it is a point of honour to try to regain, during other potlatches, each one of the quarters and rivet them together until the object is once again complete. A copper of this kind increases in value (Boas, *Sec. Soc.*, p. 334).

    In any case to expend them, or to smash them is to kill them (*Ethn. Kwa.*, p. 1285, lines 8, 9). The general expression is 'to cast them into the sea'; it is common also to the Tlingit (*Tlingit T.M.*, pp. 63, 399, song no. 43). If the copper objects do not sink and do not die, it is because they are counterfeit, made of wood, and thus come to the surface (story of a Tsimshian potlatch where the opponents were the Haïda, *Tsim. Myth.*, p. 369). When smashed up, it is said that they lie 'dead on the seashore' (Kwakiutl) (Boas, *Sec. Soc.*, p. 564, and n. 5).

245   It would seem that among the Kwakiutl there were two kinds of copper objects: the more important ones that do not go out of the family and that can only be broken to be recast, and certain others that circulate intact, that are of less value, and that seem to serve as satellites for the first kind (e.g. Boas, *Sec. Soc.*, pp. 564, 579). The possession of this secondary kind of copper object doubtless corresponds among the Kwakiutl to that of the titles of nobility and second-order ranks with whom they travel, passing from chief to chief, from family to family, between the generations and the sexes. It appears that the great titles and the great copper objects at the very least remain unchanged within the clans and tribes. Moreover, it might be difficult for it to be otherwise.

246   A Haïda myth of the potlatch of chief Hayas relates how a copper object sang: 'This thing is very bad. Stop Gomsiwa (the name both of a town and a hero); around the little copper object there are many others' (*Haïda Texts*, Jesup, vol. 6., p. 760). It refers to a 'little copper object' that becomes 'big' by itself, and around which others assemble (cf. elsewhere, the copper salmon).

247   In a child's song (*Ethn. Kwa.*, p. 1312, lines 3, 14) 'the copper objects with the names of great chiefs of the tribes will gather around him'. The copper objects are deemed to 'fall of their own accord into the household of the chief' (name of a Haïda chief, Swanton, *Haïda*,

p. 274, E). They 'meet up with one another in the house', they are 'flat things that join each other' (*Ethn. Kwa.*, p. 701).

248 See the myth of the 'Bringer of copper objects' in the myth of 'the 'Inviter' (Quexsot' onox) (*Kwa. T.*, vol. 3, p. 248, lines, 25, 26). The same copper object is called 'Bringer of properties' (Boas, *Sec. Soc.*, p. 415). The secret song of the noble who bears the title of Inviter runs:

> My name will be 'property making its way towards me', because of my 'Bringer' of properties.
> The copper objects make their way towards me because of the 'Bringer' of copper objects.

The Kwakiutl text says very precisely, 'The *aqwagila*', the 'Maker of copper objects', and not merely the 'Bringer' of them.

249 For example, in a speech at a Tlingit potlatch (*Tlingit T.M.*, p. 379); (Tsimshian) the copper object is a shield (*Tsim. Myth.*, p. 385).

250 In a speech about the giving of copper objects in honour of a son who has been recently initiated, 'the coppers given are an "armour", an "armour of property"' (Boas, *Sec. Soc.*, p. 557) – alluding to copper objects strung around the neck. The title of the young man, moreover, is Yaqois, the 'Bearer of property'.

251 An important ritual that takes place when the Kwakiutl princesses are shut away at puberty illustrates these beliefs very well; they wear copper objects and abalone shells, and at that time they themselves take the title of copper objects, of 'things flat and divine, meeting one another in the house'. It is said then that 'they and their husbands will easily acquire copper objects' (*Ethn. Kwa.*, p. 701). 'Copper objects in the house' is the title of the sister of an Awikenoq hero (*Kwa. T.*, vol. 3, p. 430). A song of the daughter of a Kwakiutl noble, foreseeing a kind of *svayamvara*, a choice by the husband, as among the Hindus, perhaps belongs to the same ritual, and is expressed thus: 'I am seated on the copper objects. My mother is weaving my belt for the day when I shall have "household dishes"' (*Ethn. Kwa.*, p. 1314).

252 The copper objects are often identical with the spirits. There is the well-known theme of the shield and the living heraldic emblem. There is identity of the copper object with both the 'Dzonoqoa' and the 'Qominoqa' (*Ethn. Kwa.*, p. 860, p. 1421). The copper objects are totem animals (Boas, *Tsim. Myth.*, p. 460). In other cases they are

only attributes of certain mythical animals. 'The copper doe' and its 'copper antler branches' play a role in the Kwakiutl summer festivals (Boas, *Sec. Soc.*, pp. 630, 631; cf. p. 729: 'greatness on its body' (literally, 'wealth on its body')). The Tsimshian consider copper objects to be like 'the hair of spirits' (Boas, *Sec. Soc.*, p. 326); like the 'excrements of spirits' (list of themes (Boas, *Tsim. Myth.*, p. 837)); and like the claws of the woman otter (ibid, p. 563). The copper objects are used by the spirits in a potlatch they give among themselves (*Tsim. Myth.*, p. 285; *Tlingit T.M.*, p. 51). The copper objects 'are pleasing to them'. For comparisons, see Boas, *Tsim. Myth.*, p. 846.

253   The song of Neqapenkem (Face of Ten elbow-lengths): 'I am pieces of copper, and the chiefs of the tribes are broken copper objects', see Boas, *Sec. Soc.*, p. 428; cf. p. 667 for the text and a literal translation.

254   Dandalayu the copper object 'groans in his house' in order to be given away (Boas, *Sec. Soc.*, p. 622 (speech)). Maxtoselem the copper object 'complained that they would not break him'. The blankets with which he is paid 'keep him warm' (Boas, *Sec. Soc.*, p. 572). One recalls that he bears the title, 'He whom the other copper objects are ashamed to look at'. Another copper object shares in the potlatch and 'is shameful' (*Ethn. Kwa.*, p. 882, line 32).

Another Haïda copper object ((Masset), *Haïda Texts*, Jesup, vol. 6, p. 689), the property of the chief, 'He whose property makes a noise', signs after having been broken: 'I shall rot here, I have dragged a large number' (into death, because of the potlatches).

255   The two rituals of the donor or the recipient buried under, or walking on the piles of blankets are equivalent. In the first case, one is superior, and in the other, inferior to his wealth.

256   A general remark: we know fairly accurately how and why, and during which ceremonies, expenditure and destruction are the means of passing on goods in the American Northwest. However, we are still badly informed about the forms assumed by the act of passing on things, particularly copper objects. This question should be the subject of an investigation. The little we know is extremely interesting and certainly denotes the link between property and its owners. Not only is what corresponds to the passing on of a copper object called: 'to put the copper object in the shadow of the name' of so-and-so, but among the Kwakiutl, its acquisition 'lends weight' to the new proprietor (Boas, *Sec. Soc.*, p. 349). Among the Haïda, not only does

one raise a copper object to signify that one is buying a parcel of land (*Haïda T.M.*, p. 86), but copper objects are also used as percussion instruments, as in Roman law: with them one strikes the people to whom they are given; this ritual is demonstrated in a story (Skidegate) (ibid, p. 432). In this case, the things touched by the copper instrument are annexed by him, are killed by him; this is, moreover, a 'peace' or 'gift' ritual.

At least in a myth, the Kwakiutl (Boas, *Sec. Soc.*, pp. 383, 385; cf. p. 677, line 10) have retained the memory of a transmission rite that is to be found among the Eskimos: the hero bites all that he is given. A Haïda myth describes how Lady Mouse 'licked' all she was given (*Haïda Texts*, Jesup, vol. 6, p. 191).

257 In a marriage rite (the breaking of a symbolic boat) they sing:

> I am going to Mount Stevens and will break it into pieces. I shall make stones of it for my fire (debris).
> I am going to Mount Qatsaî and will break it. I shall make stones of it for my fire.
> Wealth is rolling towards him, coming from the great chiefs.
> Wealth is rolling towards him from all sides.
> All the great chiefs are going to have themselves protected by him.

258 Normally, moreover, at least among the Kwakiutl, they are identical. Certain nobles are identified with their potlatch. The main title of the principal chief is even simply Maxwa, which means 'Great potlatch' (*Ethn. Kwa.*, pp. 805, 972, 976). Cf. in the same clan the names 'Givers of potlatch', etc. In another tribe of the same nation, among the Dzawadaenoqu, one of the principal titles is that of 'PoLas' (sic). See above, n. 209. See *Kwa. T.*, vol. 3, p. 43, for the genealogy. The principal chief of the Heiltsuq is in contact with the spirit 'Qominoqa', the 'Rich One', and bears the name 'Maker of Wealth' (ibid, pp. 424, 427). The Qaqtsenoqu princes have 'summer names', i.e. the names of clans that exclusively designate 'properties', names in 'yag': 'Property over the body', 'Great property', 'Having property', 'Place of property' *Kwa. T.*, vol. 3, p. 191; cf. p. 187, line 14). Another Kwakiutl tribe, the Naqoatoq, gives its chief the title of 'Maxwa' and Yaxlm', 'Potlatch' and 'Property'; this name figures in the myth of the 'Stone body' (cf. 'Ribs of stone', the son of Lady Fortune, Haïda). The spirit says to him: 'Your name will be "Property", Yaxlem' (*Kwa. T.*, vol. 3, p. 215, line 39).

In the same way, among the Haïda a chief bears the name: 'He who cannot be bought' (the copper object that the rival cannot buy) (Swanton, *Haïda*, p. 294, 16, I). The same chief also bears the title: 'All mingled together', namely, 'Potlatch gathering' (ibid, no. 4; cf. elsewhere, the titles 'Properties in the house').

3 THE SURVIVALS OF THESE PRINCIPLES IN ANCIENT SYSTEMS OF LAW AND ANCIENT ECONOMIES

1   We know, of course, that the facts have another dimension (see Ch. 4, n. 38, p. 156), and the stopping of research at this point is only temporary.

2   Meillet and Henri Lévy-Bruhl, as well as our much regretted colleague, Huvelin, have kindly given us valuable advice about the next paragraph.

3   We know that, apart from hypothetical reconstitutions of the Twelve Tables and a few legal texts preserved on inscriptions, we have only very poor sources for the whole of the first four centuries of Roman law. However, we shall not adopt the extremely critical attitude of Lambert (1906) *L'Histoire traditionnelle des Douze Tables*, (Mélanges Appleton). But we must agree that a large part of the theories of Romanist scholars, and even those of Roman antiquarians themselves, must be treated as hypotheses. We take leave to add another hypothesis to the list.

4   On Germanic law, see elsewhere.

5   On the *nexum*, see Huvelin, '*Nexum*' in *Dict. des Ant.*; 'Magie et Droit individuel', *Année Sociologique* 10, and the analyses and discussions by him in *Année Sociologique* 7: 472 ff.; 9: 412 ff.; 11: 442 ff.; 12: 482 ff.; Davy, 'Foi jurée', p. 135; for a bibliography and the theories of Romanist scholars, see Girard, *Manuel élémentaire de Droit romain*, 7th edn, p. 354.

From every viewpoint Huvelin and Girard seem to us to be very close to the truth. We suggest only one addition and one objection to Huvelin's theory. The 'insults clause' ('Magie et droit individuel', p. 28; cf. 'Injuria' (Mélanges Appleton)) is, in our opinion, not only magical. It is a very clear case, and a vestige of the ancient rights to potlatch. The fact that one is a debtor and the other a creditor enables the one who is superior in this way to insult his opponent, who is under an obligation to him. From this there flows an important series of relationships to which we draw attention in that volume of the *Année*

Sociologique, concerning 'joking relationships', especially those of the Winnehago (Sioux) tribe.

6 Huvelin, 'Magie et droit individual', *Année Sociologique* 10.

7 See p. 60; on the *wadiatio*, see Davy, *Année Sociologique* 12: 522, 523.

8 This interpretation of the word *stips* is based on that of Isidore of Seville. See vol. 5, pp. 24, 30; Huvelin (1906) *Stips, stipulatio . . . (Mélanges Fadda)*; Girard, *Manuel*, p. 507, n. 4, following Savigny, opposes the texts of Varro and Festus to this purely and simply figurative interpretation. But Festus, after in fact having said: *stipulus, firmus*, in a sentence that has unfortunately been in part destroyed, must have spoken of a '[. . . ?] *defixus*', perhaps a stick stuck in the ground (cf. the throwing of a stick at the sale of land mentioned in the contracts of the Hammurabi era in Babylon. See Cuq (1910) 'Etude sur les contrats . . . ', *Nouvelle Revue Historique du Droit*, p. 467.

9 See Huvelin, loc. cit., in *Année Sociologique* 10: 33.

10 We shall not enter into the debates between Romanist scholars. But we shall add a few remarks to those of Huvelin and Girard concerning the *nexum*: [1] The word itself derives from *nectere*. Regarding this last word, Festus (under the word's listing; cf. also heading, *obnectere*) has preserved one of the rare documents of the Pontiffs that have come down to us: *Napuras stramentis nectito*. The document clearly alludes to the property taboo, indicated by straw knots. Thus the thing *tradita* is itself marked and bound, and comes to the *accipiens* provided with this binding. It can therefore bind him. [2] The individual who becomes *nexus* is the recipient, the *accipiens*. Now, the solemn formula of the *nexum* supposes that he is *emptus*, 'bought', as it is normally translated. But *emptus* really means *acceptus*. The individual who has received the thing is himself accepted by the loan, even more than bought: this is because he has received the thing and has received the copper ingot that, as well as the thing, is given him on loan. The question is debated whether, in this operation, there is *damnatio, mancipatio*, etc. (Girard, *Manuel*, p. 503). Without taking sides in this question, we believe that all these terms are comparatively synonymous. (Cf. the expression *nexo mancipioque* and that of *emit mancipioque accepit* on inscriptions (sales of slaves).) Moreover, nothing is more simple than this synonymous connection, since the mere fact of having accepted something from someone places you under an obligation to him: *damnatus, emptus, nexus*. [3] It seems to us that Romanist scholars, and even Huvelin, have generally not paid enough attention to a detail relating to the formalism of the *nexum*:

the fate of the bronze ingot, the *aes nexum* so discussed by Festus (under the heading *nexum*). This ingot, at the time when the *nexum* is formed, is given by the *tradens* to the *accipiens*. But, we believe, when the latter discharges himself from the bond, not only does he carry out the service promised or hand over the thing or its price, but in particular, on the same pair of scales and before the same witnesses, he returns this same *aes* to the lender, to the seller, etc. Then he buys it and receives it in his turn. This ritual of the *solutio* of the *nexum* is fully described to us by Gaius, III, 174 (the text has been largely reconstituted; we adopt the reading accepted by Girard; cf. *Manuel*, p. 501, n; Cf. ibid, 751). In a cash sale the two actions occur, so to speak, at the same time, or at very short intervals. This double symbol was less apparent than in a sale on credit or in a loan that was solemnly carried out. This is why the dual interplay has not been perceived. But it functioned in the cash sale all the same. If our interpretation is exact, in addition to the *nexum* that derives from the solemn forms employed, and the *nexum* that comes from the thing, there is in fact another *nexum* that arises from this ingot that is given and accepted in turn, and weighed on the same scales, *hanc tibi libram primam postremamque*, by the two contracting parties, thus alternately linked together. [4] Moreover, let us assume for a moment that we can imagine a Roman contract before bronze money was used, or even this weighed-out ingot, or that piece of moulded copper, the *aes flatum*, that represented a cow (we know that the first Roman monies were struck by the *genies*, and, since they represented cattle, were doubtless title deeds that committed the cattle of these *gentes*). Let us suppose a sale where the price was paid in real cattle, or figuratively in them. It is sufficient to realize that the handing over of this cattle price, or its representation, brought together the contracting parties, and in particular brought the seller to the buyer. As in a sale or in any disposal of cattle, the buyer or the last possessor remains, for a time at least (because of the possibility of irregularities on the buyer's part, etc.) in contact with the seller or the previous possessor. (See elsewhere, actions in Hindu law and folklore.)

11 Varro, *De re rustica*, II, part 1, 15.
12 On *familia*, see *Digesta*, L, XVI, *De verbo sign.*, no. 195, ss. 1. '*Familiae appellatio*, etc. . . . *et in res et in persona diducitur*, . . . etc.' (Ulpianus). Cf. Isidore of Seville, XV, 9. 5. In Roman law, up to a very late period the dividing up of an inheritance was called *familiae creiscundae*, *Digesta*, XI, II. Again, in the Code, III, XXXVIII. Conversely *res = familia*;

in the Twelve Tables, V, 3, *super pecunia tutelave suae rei*. Cf. Girard, *Textes de droit romain*, p. 869 n. *Manuel*, p. 322; Cuq, *Institutions*, I, p. 37. Gaius II, 224, reproduces this text, stating *super familia pecuniaque*. *Familia* – *res* and *substantia* also in the Code (Justinian), VI, XXX, 5. Cf. also *familia rustica et urbana*, *Digesta*, L, XVI, *De verbo sign.*, no. 166.

13 Cicero, *De Oratione*, 56; *Pro Caecina*, VII. Terence: 'Decem dierum vix mihi est familia.'

14 Walde, *Lateinisches Etymologisches Wörterbuch*, p. 70. Walde hesitates about the etymology he propounds, but there need be no hesitation. Moreover, the principal *res*, the supreme *mancipium* of the *familia*, is the slave *mancipium*, whose other name of *famulus* has the same etymology as *familia*.

15 On the distinction *familia pecuniaque* attested to in the 'sacratae leges' (see Festus, under this heading), and by numerous texts, see Girard, *Textes*, p. 841, n. 2, *Manuel*, p. 263, n. 3; p. 274. It is clear that the nomenclature has not always been very certain, but, contrary to Girard's view, we believe that originally, very long ago, there was a very precise distinction. The division is, moreover, found in Osque [sic: language of people living in Latium that was at the roots of Latin], *famelo in cituo* (*Lex Bantia*, line 13).

16 The distinction between *res mancipi* and *res nec mancipi* only disappeared from Roman law in A.D. 532 by a special abrogation of Quiritary law.

17 On the *mancipatio*, see elsewhere. The fact that it had been required, or was at least legal, until so late a period proves the difficulty the *familia* had in ridding itself of the *res mancipi*.

18 On this etymology, see Walde, p. 650. Cf. *rayih*, property, precious thing, talisman; cf. Avestic *rae*, *rayyi*, with the same meanings; cf. old Irish *rath*, 'gratuitous present'.

19 The word that designates the *res* in Osque is *egmo*, cf. *Lex Bantia*, lines 6, 11. Walde links *egmo* to *egere*, 'the thing that one lacks'. It is indeed possible that the ancient languages of Italy had two corresponding words, in antithesis to one another, to designate the thing one gives and that gives pleasure, *res*, and the thing that one lacks, *egmo*, and that one waits for.

20 See elsewhere.

21 See Huvelin, *Furtum* (Mélanges Gerard), pp. 159–75; *Etude sur le Furtum*, 1. *Les sources*, p. 272.

22 The expression of a very old law, the *lex Atinia*, preserved by Aulus Gellus, XVII, 7, '*Quod subruptum erit ejus rei aeterna auctoritas esto*'.

Cf. *Extracts from Ulpianus*, III, pp. 4, 6.; cf. Huvelin, 'Magie et Droit individuel', *Année Sociologique* 10: 19.

23 Among the Haïda, the person from whom something has been stolen has only to place a dish outside the thief's door for it normally to be returned.

24 Girard, *Manuel*, p. 265. Cf. *Digesta*, XIX, IV, *De permut.*, 1, 2: '*permutatio autem ex re tradita initium obligationi praebet*'.

25 Mod. *Regul.*, in *Digesta*, XLIV, VII, *De Obl. et act.* 52, '*re obligamur cum res ipsa intercedit*'.

26 Justinian (in A.D. 532), *Code*, VIII, LVI – 10.

27 Girard, *Manuel*, p. 308.

28 Paul, *Digesta*, XLI, I – 31, 1.

29 *Code*, II, III, *De pactis*, 20.

30 On the meaning of *reus*, 'guilty', 'responsible', see Mommsen, *Römisches Strafrecht*, 3rd edn, p. 189. The classical interpretation springs from a kind of historical *a priori* that makes personal, and in particular, criminal public law, the primitive law, and which sees in property rights and contracts modern, sophisticated phenomena. Whereas it would be so simple to deduce the rights of contract from the contract itself!

31 *Reus*, moreover, belongs to the language of religion (see Wissowa, *Religion und Kultus der Römer*, p. 320, notes 3, 4) no less than to that of the law: *voti reus*, *Aeneid* V, 237; *reus qui voto se numinibus obligat* (Servius ad *Aeneid*, IV, 699). The equivalent of *reus* is *voti domnatus* (Virgil, *Eclogues*, V, 80); and this indeed symptomatic, since *damnatus* = *nexus*. The individual who has made a vow is in exactly the same position as one who has promised or received a thing. He is *damnatus* until he has been acquitted.

32 *Indo-germanische Forschungen*, XIV, p. 131.

33 *Lateinisches Etymologisches Wörterbuch*, p. 651, cf. *reus*.

34 It is the interpretation of the oldest Roman jurists themselves (Cicero, *De Or.*, II, 183, '*Rei omnes quorum de re disceptatur*'); they had always present in their minds the meaning *res*, 'lawsuit'. But this is interesting in that it keeps the memory alive of the time of the Twelve Tables, II, 2, in which *reus* not only designates the person indicted but the two parties in the whole affair, the *actor* and the *reus* of recent procedures. Festus (under *reus*, cf. another fragment, '*pro utroque ponitur*'), commenting upon the Twelve Tables, cites two very old Roman lawyers upon this subject. Cf. Ulpianus in *Digesta*, II, IX, 2, 3, *alteruter ex litigatoribus*. The two parties are both equally bound by the suit. There

is reason to suppose that they were equally bound by the thing beforehand.

35 The notion of *reus*, responsible for a thing, made responsible by the thing, is still familiar to the very old Roman lawyers that Festus cites: '*reus stipulando est'idem qui stipulator dicitur . . . reus promittendo qui suo nomine alteri quid promisit*,' etc. Festus evidently is referring to the modification in the meaning of these words in the system of caution contract that is termed correality. But the older authors were speaking of something different. Moreover, correality (Ulpianus in *Digesta*, XIV, VI − 7, 1 and the title *Dig*. XLV, II, *de duo reis const.*), has kept the meaning of that indissoluble link that binds the individual to the thing, in point of fact, the affair, and with it, 'his friends and parents', who are correal partners.

36 In the *Lex Bantia*, in Osque, '*ministreis – minoris partis*' (line 19), is the party that fails in the suit, This shows how the meaning of these terms has never been lost in the dialects of Italy.

37 Romanist scholars seem to date the division between *mancipio* and *emptio venditio* too far back. At the time of the Twelve Tables and probably long afterwards, it is highly unlikely that there were sale contracts that were pure contracts by agreement, as they became later at a date that we can roughly surmise, some time in the era of Scaevola. The Twelve Tables use the term *venum duuit* only to designate the most solemn form of sale that can be carried out, that of a son – one which certainly could not be operated except through *mancipatio* (Twelve Tables, IV, 2). Moreover, at least for matters relating to *mancipi*, at that time a sale was performed exclusively, as a contract, by a *mancipatio*. All these terms are therefore synonymous. The Ancients had some memory of this confusion (see Pomponius, *Digesta*, XL, VII, *de statuliberis: 'quoniam* Lex XII. T. *emotionis verbo omnem alienationem complexa videatur'*). Conversely, the word *mancipatio* for a very long time, up to the Actions of the Law, designated acts that are pure contracts by agreement, such as the *fiducia*, with which it is sometimes confused. See documents in Girard, *Manuel*, p. 545 (cf. p. 299). Even *mancipatio*, *mancipium*, and *nexum* were, doubtless from a very early time, used without much distinction between one another.

However, whilst retaining this synonymous relationship, in what follows we consider exclusively the *mancipatio* of *res* that form part of the *familia*, and we base ourselves on the principle preserved in Ulpianus, XIX, 3 (cf. Girard, *Manuel*, p. 303: '*mancipatio . . . propria alienato rerum mancipi*').

38 For Varro, *De re rustica*, II, 1, 15; II, 2, 5; II, V, 11; II, 10, 4, the word *emptio* includes *mancipatio*.

39 One can even imagine that this *traditio* was accompanied by rituals of the kind like those preserved for us in the formalism of *manumissio*, of the freeing of the slave who is considered to be buying himself. We are ill-informed about the actions of the two parties in the *mancipatio*; on the other hand it is truly remarkable that the formula for *manumissio* (Festus, under the heading *puri*) is, all in all, identical to that of the *emptio venditio* of cattle. Perhaps, after having taken into his hand the thing he was handing over, the *tradens* struck it with the palm of his hand. One can compare the *vus rave*, the tap on the pig (Bank Islands, Melanesia) and, in our own cattle markets, the tap on the hindquarters of the cattle sold. However, these are hypotheses that we would not allow ourselves to entertain unless the texts, and particularly that of Gaius, were not, at this very point, full of gaps that the discovery of other manuscripts will one day doubtless remedy.

Let us also bear in mind that we have found a formalism identical to that of the 'percussion' of the emblazoned copper object among the Haïda. See Ch. 2, n. 256, p. 136.

40 See above, our remarks on the *nexum*.

41 Cuq, *Institutions juridiques des Romains*, vol. 2, p 454.

42 See above. The *stipulatio*, the exchange of the two fragments of the stick, not only corresponds to ancient pledges, but to ancient, additional gifts.

43 Festus (*manumissio*).

44 See Varro, *De re rustica* 2, 1, 15; 2, 5; 2, 5, 11: *sanos, noxis, solutos*, etc.

45 Note also the expressions *mutui datio*, etc. In fact the Romans had no other word than *dare*, to give, to designate all the acts that made up the *traditio*.

46 Walde, ibid, p. 253.

47 *Digesta*, XVIII, I–33, extracts by Paul.

48 On words of this type, see Ernout (1911) *Credo-Craddhâ* (Mélanges Sylvain Lévi). This is another case of identity, as for *res* and so many other words, between the Italo-Celtic and Indo-Iranian legal vocabularies. We note the archaic forms of all these words: *tradere, reddere*.

49 See Walde (ibid) under the heading *vendere*.

It is even possible that the very old term of *licitatio* has still about it a memory of the equivalence of war and sale (by auction): '*Licitati in mercato sive pugnando contendentes*', says Festus (under the word, *licitati*). Compare the Tlingit and Kwakiutl expression: 'war of prop-erty', and Ch. 2, n. 147, p. 114, for auctions and potlatches.

50 We have not sufficiently studied Greek law, or rather the surviving fragments of the law that must have preceded the major codifications carried out by the Ionians and the Dorians, to be able to say whether the various Greek peoples were ignorant of, or knew these rules con-cerning the gift. It would be necessary to review a complete literature dealing with various questions: gifts, marriages, pledges (see Gernet (1917) 'Ἐγγύχί', *Revue des études grecques*: cf. Vinogradoff, *Outlines of the History of Jurisprudence*, vol. 2, p. 235), hospitality, interest, and contracts, and we would even only find fragments of them. However, there is one in Aristotle (*Nichomachean Ethics*, 1123 a. 3), concerning the magnanimous citizen and his public and private expenditure, his duties and burdens, which mentions the reception of foreigners, the embassies χαί δωρεάζ χαί ἀυτιδρεάξ, how they spend εἴζ τά χοινά, and he adds: τά δὲδῶρα τοῖε ἀναθνασίν ἐγει τι ὅμοιν. 'Gifts have some-thing analogous to consecrations about them' (cf. Ch. 2, n. 152, p. 116, Tsimshian).

Two other systems of living Indo-European law, the Albanian and the Ossetian, present institutions of this kind. We shall confine our-selves to referring to the laws or modern decrees that forbid or limit among these peoples overlavish expenditure on marriages or deaths, etc.; e.g. Kovalewski, *Coutume contemporaine et Loi ancienne*, p. 187, n.

51 We know that almost all the formulas of contract employed are attested to in the Aramaic papyri of the Philae Jews in Egypt in the fifth century B.C. (see Cowley (1923) *Aramaic Papyri*, Oxford). We are also aware of studies by Ungnad of Babylonian contracts (see Huvelin, *Année Sociologique* 12: 508, and Cuq (1910) 'Etudes sur les contrats de l'époque de la Ire Dynastie babylonienne', *Nouvelle Revue de l'Histoire du Droit*).

52 Ancient Hindu law is known to us through two series of collections drawn up fairly late in comparison with the rest of the [Hindu] 'Scrip-tures'. The most ancient series consists of the *Dharmasutra*, to which Bühler assigns a date before Buddhism ('Sacred Laws', in *Sacred Books of the East*, Introduction). But it is not clear whether a certain number of these *sutra* – if not the tradition on which they are based – do not date from after Buddhism. In any case, they form part of what the Hindus call the *Çruti*, the Revelation. The other series is that of the *smrti*, the Tradition, or the *Darmaçastra*: Books of the Law, the

principal one of which is the celebrated Code of Manu, which, for its part, comes scarcely later than the *sutra*.

However, we have preferred to use a long epic document that, in the Brahmin tradition, has a *smirti* and *çastra* value (tradition and taught law). The *Anuçasanaparvan* (Book XIII of the *Mahabharata*) is explicit in a totally different way from the books of the law concerning the morality of the gift. Moreover, it has as much value and possesses the same inspiration as these latter do. In particular, it appears that its composition is based on the same tradition of the Brahmin school of the Manava as that on which the Code of Manu itself relies (see Bühler, 'The Laws of Manu', in *Sacred Books of the East*, p. 70 ff.). Moreover, one might say that this *parvan* and Manu quote from each other.

In any case this latter document is of inestimable value. It is an enormous book about an enormous epic of the gift, *dana-dharmakathanam*, as the commentary says, to which more than one-third of the book – more than forty 'readings' – are devoted. Moreover, this book is extremely popular in India. The poem tells how it was recited in tragic fashion to Yudhisthira, the great king, the incarnation of Dharma, the Law, by the great King and Seer Bhisma, lying on his bed of nails, at the moment of his death.

We shall refer to it from now on as: *Anuç.*, and generally will give the two references: the number of the line, and the number of the line by *adhyaya*. The characters used in the transcription are replaced by italics.

53 It is very clear, from more than one characteristic, that, if not the rules, at least the versions drawn up of the *çastra* and the epics, date from after the struggle against Buddhism that they recount. In any case this is certainly so for the *Anuçasanaparvan*, which is full of allusions to that religion (see especially the *Adhyaya* 120). Since the date of drawing up the definitive versions may be so late, one might perhaps find an allusion to Christianity, precisely in relation to the theory of gifts, in the same *parvan* (*Adhyaya*, 114, line 10), where Vyasa adds: 'This is the law taught with subtlety (*nipunena*, Calcutta), (*naipunena*, Bombay): 'Let him not do to others what is contrary to his own self, this is the summation of the *dharma* (the law)' (line 5673). Yet on the other hand it is not impossible for the Brahmins, those coiners of formulas and proverbs, to have arrived at such an adage by themselves. In fact, the preceding line (line 9 = 5672) has a profoundly Brahminic ring to it, 'Another lets himself be guided by desire (and deceives himself). In

rejection and the gift, in good and bad fortune, in pleasure and lack of pleasure, it is in relation to himself (to his self) that man measures them (the things), etc.' The commentary of Nilakantha is formal, very original, and pre-Christian: 'As someone conducts himself towards others, so do others conduct themselves towards him. It is by feeling how one would oneself receive a refusal after having made a request . . . etc. . . . that one sees how one must give.'

54 We do not mean that, from a very early time, that of the writing of the *Rg Veda*, the Aryans who had arrived in Northeast India did not know about the market, the trader, price, money, and sale (see Zimmern, *Altindisches Leben*, p. 257 ff.): *Rg Veda* IV, 24, 9. Above all, the *Atharva Veda* is familiar with this form of economy. Indra himself is a merchant. (Hymn, III, 15, used in *Kauçika-sutra*, VII, 1, VII, 10 and 12, in a ritual about a man going to a sale. See, however, *dhanada*, ibid, line 1 and *vajin*, the epithet of Indra, ibid).

Nor do we mean that contract in India had solely this origin: 'real' party, personal party, and formal party in the transmission of goods; nor that India has not known other forms of obligation, for example, the quasi-offence. We seek only to demonstrate the fact of the survival alongside these forms of law, of another form, another economy, and another mentality.

55 In particular, there must have been – as there are still among aboriginal tribes and nations – total exchange of services in clans and villages. The prohibition laid upon the Brahmins (*Vasistha*, 14, 10 and *Gautama*, XIII, 17; *Manu*, IV, 217) not to accept anything at all from 'the multitudes', and above all not to participate in any feast given by them, is surely aimed at usages of this kind.

56 *Anuç.*, lines 5051, 5045 (= *Adh.* 104, lines 95, 98): 'Let him not consume any liquid whose essence has been removed . . . nor without having passed it as a gift to the one who is seated at table with him' (commentary: 'and the one he has caused to sit down with him, and who must eat with him').

57 For example, the *adanam*, the gift that friends make to the parents of the young man who has been tonsured or of the young initiate, or to the engaged couple, etc., is identical, even as regards the title, with the Germanic *Gaben* that we mention later (see the *grhyasutra* (domestic rituals), Oldenberg, *Sacred Books*, Index, under these various headings).

There is another example: the honour that comes from gifts (of food), *Anuç.*, 122, lines 12, 13, 14: 'honoured, they do honour;

decorated, they decorate others. "Here and there," it is said, "is a giver", and he is glorified on all sides' (*Anuç.*, line 5850).

58 An etymological and semantic study would, moreover, allow us to arrive at conclusions analogous with those regarding Roman law. The oldest Vedic documents bristle with words whose etymologies are even clearer than those of the Latin terms, and that all suppose, even those that concern the market and selling, another system in which exchanges, gift, and wagers took the place of the contracts we normally think of when we speak about these things. The uncertainty (moreover, one general in all Indo-European languages) has often been noted concerning the meaning of the Sanskrit word *da* (and its infinitely numerous derivatives) that we translate as 'to give', e.g. *ada*, 'to receive', 'to take', etc.

Let us choose as another example the two Vedic words that best designate the technical act of sale. They are: *parada çulkaya*, to sell at a price, and all the words derived from the verb *pan*, e.g. *pani*, merchant. Apart from the fact that *parada* includes *da*, to give, *çulka*, which has really the technical sense of the Latin *pretium*, means something completely different: it signifies not only value and price, but also: the price of the fight, the bride-price, payment for sexual services, taxes, and tribute. The verb *pan* has, from the time of the *Rg Veda*, given the word *pani* (merchant, miserly, cupidous, and a name for foreigners). The name for money, *pana* (later, the well-known *karsapana*), etc. means to sell, as well as to play, to wager, to fight for something, to give, to exchange, to risk, to dare, to win, to bring into play. Moreover, we undoubtedly do not need to suppose that *pan*, 'to honour, to praise, to appreciate', is a different verb from the first one. *Pana* (money) also means: the thing one sells, the wages, the object of the wager and the game, the gaming-house, and even the inn that replaces hospitality. All this vocabulary links ideas that only come together in the potlatch. Everything points to the original system that was used to conceive the later system of selling proper. However, let us not pursue this attempt at reconstruction by means of etymology. It is not necessary in the case of India and would lead us far, doubtless beyond the Indo-European world.

59 See résumé of the epic in *Mhbh. Adiparvan*, reading 6.

60 See, for example, the legend of Hariçcandra, *Sabhaparvan, Mahbh.*, Book 2, reading 12. Other examples in *Virata Parvan*, reading 72.

61 Concerning the main subject of our analysis, the obligation to reciprocate, we must acknowledge that we have found few facts in

Hindu law, except perhaps *Manu*, VIII, 213. Even so, the most apparent fact is the rule that forbids reciprocity. Clearly, it seems that originally the funeral *çraddha*, the feast of the dead that the Brahmins expanded so much, was an opportunity to invite oneself and to repay invitations. But it is formally forbidden to act in this way, for example, *Anuç.*, lines 4311, 4315 = XIII, reading 90, lines 43 ff.: 'He who invites only friends to the *çraddha* does not go to heaven. One must not invite friends or enemies, but neutral persons, etc. The remuneration of the priests offered to priests who are friends is called demoniacal (*picaca*)' (line 4316). This prohibition doubtless constitutes a real revolution as compared with the then current customs. Even the lawyer poet links it to a certain time and a certain school (*Vaikhanasa Çruti*, ibid, line 4323 = reading 90, line 51). The cunning Brahmins in fact entrusted the gods and the shades with the task of returning gifts that had been made to themselves. Undoubtedly, the common mortal continued to invite his friends to the funeral meal. Moreover, this continues in India to the present day. For his part, the Brahmin did not return gifts, did not invite, and did not even, all said and done, accept invitations. However, Brahmin codes have been preserved in sufficient documents to illustrate our case.

62 *Vas. Dh. su.*, XXIX, 1, 8, 9, 11–19 = *Manu*, IV, 229 ff. Cf. *Anuç.*, all readings from 64–9 (with quotations from the *Paraçara*). All this part of the book seems to be based on a kind of litany; it is half astrological and begins with a *danakalpa*, reading 64, setting out the constellations under which this or that must be given by someone to another person.

63 *Anuç.*, 3212; even that which is offered to the dogs and to the *çudra*, to 'the one who cooks for the dog', [*susqui* cooks the dog] *çvapaka* (= reading 63, line 13; cf. ibid, line 45 = lines 3243, 3248).

64 See the general principles concerning the way, in the series of rebirths, one finds once more the things given (XIII, reading 145, lines 1–8, lines 29, 30). The sanctions relating to the miser are set out in this same reading, lines 15–23. In particular, he is 'reborn into a poor family'.

65 *Anuç.*, 3135. Cf. 3162 (= reading 62, lines 33, 90).

66 Line 3162 (= ibid, line 90).

67 In the end all this *parvan*, this song of the *Mahabharata*, is an answer to the following question: how can one possess Çri (Fortune), the fickle goddess? A first answer is that Çri lives among the cows, in their dung and urine, where the cows, those goddesses, have allowed it to

reside. This is why to make a gift of a cow gives an assurance of happiness (reading 82; see below, n. 79.) A second answer that is basically Hindu, and is even the basis for all the moral doctrines of India, teaches that the secret of fortune and happiness (reading 163) is to give, not to keep, not to look for fortune, but to distribute it, so that it may come back to you in this world of its own accord, and in the shape of the good you have done, as well as in the other world. To renounce the self, to acquire only to give, this is the law of nature and the source of true profit (line 5657 = reading 112, line 27): 'Each one of us must make his days fertile by distributing food.'

68 Line 3136 (= reading 62, line 34) calls this stanza a *gâtha*. It is not a *çloka*; it comes, therefore, from an ancient tradition. Moreover, I believe that the first half-line *mamevadaitha, mam dattha, mam dativa mamevapsyaya* (line 3137 = reading 62, line 35) may very well be isolated from the second half. Moreover, line 3132 isolates it beforehand (= reading 62, line 30): 'As a cow runs towards its calf, its full teats dropping milk, thus the blessed land runs towards the giver of lands.'

69 *Baudhayana Dh. su.*, 11, 18, – clearly contemporaneous not only with these rules of hospitality, but also with the Cult of Food, which can be said to be contemporaneous with the later forms of Vedic religion and which lasted until Vishnuism, into which it was integrated.

70 Brahmin sacrifices of the late Vedic era; cf. *Baudh. Dh. su.*, 11; 6, 41, 42; cf. *Taittiriya Aranyaka*, VII, 2.

71 The whole theory is expounded in the famous discussion between the *rsi* Maitreya and Tyasa, the incarnation of Krsna dvaipayana himself (*Anuç.*, XIII, 120, 121). All this discussion, in which we have found traces of the struggle of Brahminism against Buddhism (see particularly line 5802 (= XIII, 120, line 10)) must have had an historical bearing and alluded to a era in which Krishnaism triumphed. But the doctrine that is taught is indeed that of the ancient Brahmin theology and perhaps even that of the most ancient national morality of pre-Aryan India.

72 Ibid, line 5831 (= reading 121, line 11).

73 Ibid, line 5832 (= reading 121, line 12). Following the Calcutta edition, one should read it as *annam* and not *artham* (Bombay edition). The second half-line is obscure, and doubtless badly transcribed. Yet it has some meaning. 'This food that he eats, in so far as it is a food, he is its murderer, who is killed in his ignorance.' The two following lines are still enigmatic, but express more clearly the idea and allude to a doctrine that must have had a name, that of a *rsi*: line 5834 = ibid, 14), 'the

wise man, the learned man, eating the food, causes it to be reborn, he is the master – and in its turn, the food causes him to be reborn' (5863). 'This is the process of development (of things). For what is merit in the giver is merit in the receiver (and vice versa), for here there is not solely one wheel turning on a single side.' The translation of *Pratap* (*Mahabharata*) is very much a paraphrase, but it is based here on excellent commentaries and would deserve to be translated (except for an error that spoils it: *evam janayati*, line 14: it is the food and not the progeniture that is re-procreated). Cf. = *Ap. dh. su.*, 11,7,3: 'He who eats before his host destroys the food, the property, the descendants, the cattle, and the merit of his family.'

74 See above, n. 64

75 *Atharvaveda*, see 18, 3; cf. ibid, lines 19, 10.

76 I, 5, 16 (cf. above, the *aeterna auctoritas* of the thing stolen).

77 Reading 70. It concerns the gift of cows (the ritual for which is given in reading 69).

78 Line 14 ff. 'The property of the Brahmin kills, as the Brahmin's cow (kills) Nrga', line 3462 (= ibid, 33). (Cf. 3519 = reading 71, line 36.)

79 *Anuç.*, readings 77, 72; reading 76. These rules are related with a wealth of detail that is somewhat incredible and surely theoretical. The ritual is attributed to a particular school, that of Brhaspati (reading 76). It lasts three days: three nights before the act and three days afterwards; in certain circumstances, it even lasts ten days (line 3532 = reading 71, 49: line 3597 = 73, 40; 3517 = 71, 32).

80 He lived with a constant 'giving of cows' (*gavam pradana*) (line 3695 = reading 76, line 30).

81 Here what happens is an initiation of the cows to the donor and of the donor to the cows; it is a kind of mystery, *upanitesu gosu*, line 3667 (= 76, line 2).

82 At the same time it is a purification rite. In this way he delivers himself from all sin (line 3673 = reading 76, line 8).

83 Samanga (having all its limbs), Bahula (broad and fat), line 3670 (cf. line 6042, the cows said: 'Bahula, Samanga. You are without fear, you are assuaged, you are a good friend'). The epic does not fail to mention that these names are those of the Veda, of the Çruti. The sacred names are in fact also to be found in *Atharvaveda*, see 4, 18, lines 3, 4.

84 Literally, 'The giver of you, I am the giver of myself'.

85 'The action of taking [someone]': the word is exactly the equivalent of *accipere*, λχμσανειν, 'take', etc.

86 The ritual lays down that one can offer: 'cows in the form of sesame

cake or rancid butter, and also cows in gold, in silver'. In this case
they are treated as real cows, cf. 3523, 3839. The ritual, particularly
that relating to the transaction proper, is then a little more polished.
Ritual names are given to these cows. One of them means 'the
future' [cow]. The time spent with the cows, 'the cows' desire', is
then made worse.

87 *Ap. dh. su.*, 1, 17, 14; *Manu*, X, 86–95. The Brahmin can sell what has
    not been bought. Cf. *Ap. dh. su.*, 1, 19, 11.

88 Cf. Ch. 1, n. 37, p. 91, Ch. 2, n. 20, Melanesia, Polynesia; p. 99 below,
    n. 120, *Ap. dh. su.*, 1, 18, 1; *Gautama Dh. su.*, XVII, 3.

89 Cf. *Anuç.*, readings 93, 94.

90 *Ap. dh. su.*, 1, 19, 13; 3, in which Kanva, the other Brahmin school, is
    quoted.

91 *Manu*, IV, p. 233.

92 *Gautama dh. su.*, XVII, 6, 7. *Manu*, IV, 253. The list of people from
    which the Brahmin cannot accept anything (*Gautama*, XVII, 17, cf.
    Manu, IV, 215–17).

93 The list of things that must be refused, *Ap.*, 1, 18; *Gautama*, XVII. Cf.
    Manu, IV, 247–250.

94 See the whole of reading 136 of the *Anuç.* Cf. *Manu*, IV, p. 250; X, pp.
    101, 102; *Ap. dh. su.*, 1, 18, 5–8; 14–15; *Gaut.*, VII, 4, 5.

95 *Baud, dh. su.*, 11, 5, 8; IV, 2, 5: the recitation of the Taratsamandi = *Rg
    Veda*, IX, 58.

96 'The energy and the glory of the wise men is diminished because of
    the fact that they receive' (accept, take). 'Against those who do not
    wish to accept, guard yourself, O king' (*Anuç.*, line 2164 = reading 35,
    line 34).

97 *Gautama*, XVII, 19, 12 ff; *Ap.*, I, 17, 2. Formula of the etiquette of the
    gift, *Manu*, VII, p. 86.

98 *Krodho hanti yad danam*, 'Anger kills the gift', *Anuç.*, 3638 = reading
    75, line 16.

99 *Ap.*, II, 6, 19; cf. *Manu*, III, 5, 8, with an absurd theological interpret-
    ation: in this case, 'one eats the error of one's host'. This interpret-
    ation relates to the general prohibition that the laws imposed upon
    the Brahmins against exercising one of their essential occupations,
    which they still exercise, and are not supposed to do: to be the
    'eaters of sins'. In any case this means that nothing good comes
    from the gift, for any of the contracting parties.

100 One is reborn in the other world with the nature of those from whom
    one has accepted food, or of those whose food one has in the
    stomach, or of the food itself.

101 The whole theory is summed up in a reading that seems to be recent – *Anuç.*, 131, under the explicit title of *danadharma* (line 3 = 6278): 'What gifts, to whom, when, and by whom?' In this way the five motives for the gift are neatly set out: duty, when one gives spontaneously to the Brahmins; self-interest ('he gives to me, he has given to me, he will give to me'); fear ('I am not tied to him, he is not tied to me, he could do me harm'); love ('he is dear to me, I am dear to him'), 'and he gives without delay'; pity ('he is poor, and is satisfied with very little'). See also, reading 37.

102 It would also be rewarding to study the ritual whereby the thing that is given is purified, but is clearly also a means of detaching it from the giver. Water is sprinkled on it with a blade of *kuça* grass (for food, see *Gaut.*, V, 21, 18, 19, *Ap.*, II, 9, 8). Cf. the water that purifies the debt, *Anuç.*, reading 69, line 21, and commentaries of Pratap (ad locum, p. 313).

103 Line 5834; see above, n. 73

104 The facts are known from fairly late *monumenta*. The composition of the Edda songs is very much later than the conversion of the Scandinavians to Christianity. But, first, the age of the tradition can be very different from that of its composition; second, even the age of the form of the tradition that has been known the longest may be very different from that of the institution. These are two principles of criticism that the critic must never lose sight of.

In the event, one is in no danger in using these facts. A part of the gifts that occupy so prominent a place in the law we are describing are among the first institutions of the Germans of which we have evidence. It was Tacitus himself who described for us the two sorts of gifts: gifts through marriage, and the way in which they return to the family of the givers (*Germania*, XVIII, in a short chapter, to which we may return); and noble gifts, above all those of the chief, or those made to the chiefs (*Germania*, XV). Furthermore, if these customs have been preserved long enough for us to be able to uncover such traces, it was because they were solidly based and had put down strong roots in the Germanic character.

105 See Schrader and the references he indicates: *Reallexikon der indogermanischen Altertumskunde*, under the headings: *Marki*, and *Kauf*.

106 We know that the word *Kauf* and all its derivatives come from the Latin word *caupo*, 'merchant'. The uncertainty in the meaning of certain words such as *leihen, lehnen, Lohn, bürgen, borgen*, etc. is well known and demonstrates that their technical use is recent.

107 We shall not raise here the question of the *geschlossene Hauswirt-schaft*, the closed economy, of Bücher, *Entstehung der Volkswirtschaft*. From our viewpoint, the problem is badly explained. As soon as two clans existed in a society, they necessarily contracted and exchanged between one another not only their women (exogamy) and their rituals, but also their goods, at least at certain times of the year and on certain special occasions. For the rest of the time the family, often very small, lived an isolated existence. But there was never a time when it always lived in this way.

108 See these words in Kluge, and in the other etymological dictionaries of the different Germanic languages. See Von Amira on *Abgabe, Ausgabe, Morgengabe* (*Handbuch* of Hermann Paul (pages cited in the Index)).

109 The best works are still those of J. Grimm, 'Schenken und Geben', *Kleine Schriften*, II, p. 174; Brunner, *Deutsche Rechtsbegriffe besch. Eigentum*. See also Grimm, *Deutsche Rechtsalterthümer*, I, p. 246, cf. p. 297, on *Bete = Gabe*. The hypothesis that one allegedly passed from the unconditional gift to an obligatory gift is useless. There have always been the two sorts of gifts, and especially in Germanic law the two characteristics have always been mixed up together.

110 'Zur Geschichte des Schenkens', Steinhausen, *Zeitschrift für Kulturgeschichte*, see p. 18 ff.

111 See Em. Meyer, *Deutsche Volkskunde*, pp. 115, 168, 181, 183, etc. All the textbooks on Germanic folklore (Wuttke, etc.) can be consulted on this question.

112 Here we find another answer to the question posed (see Ch. 1, n. 58, p. 94) by Van Ossenbruggen concerning the magical and legal nature of the bride-price. See regarding this the remarkable theory of the relationships between the various services rendered in Morocco to the married couple, and by them, in Westermarck, *Marriage ceremonies in Morocco*, p. 361 ff., and the sections of the book that are cited in it.

113 In what follows we are not confusing the pledges with 'deposits', although the latter, of Semitic origin – as the name in Greek and Latin indicates – were known in recent Germanic law, as in our own. In certain usages they have even been confused with the ancient gifts. For example, *Handgeld* becomes *Harren* in certain Tyrol dialects.

   We omit also to demonstrate the importance of the notion of pledge as regards marriage. We only note that in Germanic dialects the 'purchase price' bears the various names of *Pfand, Wetten, Trugge,* and *Ehethaler.*

114 *Année Sociologique* 12: 29 ff. Cf. Kovalewski, 'Coutume contemporaine et loi ancienne', p. 111 ff.

115 On the Germanic *wadium* one may also consult: Thévenin, 'Contribution à l'étude du droit germanique', *Nouvelle Revue Historique du Droit,* IV: 72; Grimm, *Deutsche Rechtsalt.,* I, pp. 209–13; Von Amira, *Obligationen-Recht*; Von Amira, in *Handbuch,* Hermann Paul, I, pp. 248, 254.

   On the *wadiatio,* cf. Davy, *Année Sociologique* 12: 522 ff.

116 Huvelin, p. 31.

117 Brissaud (1904) *Manuel d'Histoire du Droit français,* p. 1381.

118 Huvelin, p. 31, n. 4 interprets this fact exclusively as being due to a degenerescence of the primitive magic ritual that may have become a mere moral theme. But this interpretation is partial and useless (see Ch. 2, n. 146, p. 114) and does not exclude the one we are putting forward.

119 To the relationship between the words *Wette,* and 'wedding', we may return later. The double ambiguity of the wager and the contract is even marked in our own languages (cf. Fr. *se défier,* and *défier*).

120 Huvelin, p. 36, n. 4.

121 On the *festucata notata,* see Heusler, *Institutionen,* I., p. 76 ff.; Huvelin, p. 33, seems to us to have neglected the use of tallies.

122 *Gift, gift. Mélanges Charles Andler,* Strasbourg, 1924. We have been asked why we have not examined the etymology of the word *gift,* the translation of the Latin *dosis,* itself a transcription of the Greek δόσιζ, 'dose, dose of poison'. This etymology presumes that High and Low German dialects would have preserved a learned name for a thing in common use. This is not the usual law in semantics. Moreover, one would need to explain the choice of the word *gift* for this translation, as well as the converse linguistic taboo that has hung over the meaning of *gift* for this word in certain Germanic languages. Finally, the Latin, and above all the Greek use of the word *dosis,* with the meaning of poison, proves that, with the Ancients, too, there was an association of ideas and moral rules of the kind that we are describing.

   We have compared the uncertain meaning of *gift* with that of the Latin word *venenum,* and those of φιλτροβ and φαρναζοβ; to this must be added the comparison (Bréal, *Mélanges de la société linguistique*, vol. 3, p. 410) of *venia, venus, venenum*, from *vanati* (Sanskrit, 'to give pleasure'), and *gewinnen*,' to win'.

An error in the quotation must also be corrected. Aulus Gellus did indeed discuss these words, but it is not he who cites Homer (*Odyssey*, IV, p. 226): it is Gaius, the jurist himself, in his book on the Twelve Tables (*Digesta*, L, XVI, *De verb. signif.*, 236).

123 *Reginsmal*, 7. The Gods have killed Otr, the son of Hreidmar. They have been forced to redeem themselves by covering the body of Otr with heaped-up piles of gold. But the god Loki curses this gold, and Hreidmar replies in the stanza quoted. We owe this fact to Maurice Cahen, who notes concerning line 3 that 'with a kind heart' is the classical translation, but *af heilom hug*, really means: 'in a frame of mind that brings good luck'.

124 This study, entitled 'Le Suicide du chef Gaulois', with notes by Hubert, will appear in a forthcoming number of the *Revue Celtique*.

125 The Chinese law relating to immovable property, like Germanic law, and our own ancient law, recognizes both the 'buy-back' form of sale, and the rights that relatives possess – very broadly defined – to buy back immovable property that has been sold but should not have been alienated from the inheritance – what is termed 'hereditary withdrawal'. See Hoang (1897) (*Variétés sinologiques*), *Notions techniques sur la propriété en Chine*, pp. 8, 9. But we do not place too much reliance on this fact: definitive sale of land is, in human history and especially in China, something so very recent. It was, right up to Roman law, and then in our ancient German and French legal systems, hedged in with so many restrictions, arising from the domestic form of communism and the deep attachment of the family to the land, and the land to the family, that the proof would have been too facile. Since the family is the home and the land, it is normal that the land should be exempt from the law and the economy of capital. In fact, the old and new laws relating to the 'homestead' [sic: in English] and the more recent French laws on 'inviolate family property' are a survival of, and a return to, the ancient situation.

126 See Hoang, ibid, pp. 10, 109, 133. I owe these facts to the kindness of Mestre and Granet, who themselves uncovered them in China.

127 *Origin and Development of Moral Ideas*, vol. 1, p. 594. Westermarck felt that there was a problem of the kind we are dealing with, but only treated it from the viewpoint of the right to hospitality. However, it is necessary to read his very important remarks on the Moroccan

custom of the *Dar* (a sacrifice that places constraint upon the sup-
plicant, ibid, p. 386) and on the principle, 'God and food will pay
him' (expressions that are remarkably identical to those in Hindu
law). See Westermarck, *Marriage Ceremonies in Morocco*, p. 365; cf.
Anthr. Ess. E. B. Tylor, p. 373 ff.

4 CONCLUSION

1 *Essays*, 2nd Series, V.
2 Cf. Koran, Sourate II, 265; cf. Kohler, in *Jewish Encyclopedia*, vol. 1,
p. 465.
3 William James, *Principles of Psychology*, vol. 2, p. 409.
4 Kruyt, *Koopen*, etc. quotes facts of this kind from the Celebes, p. 12 of
the extract. Cf. 'De Toradja's . . . ', *Tijd. v. Kon. Batav. Ges.* 43, 2, p. 299.
The rite of the introduction of the buffalo into the stable; p. 296, the
ritual of the purchase of a dog that is bought limb by limb, one part of
the body after another, and in whose food one spits; p. 281, the cat
who under no pretext is sold, but is lent out, etc.
5 This law is not based on the principle of the illegitimate nature of the
profits made by its successive owners. The law is little applied.

  Soviet legislation on literary property and its variations is very curious to study from this viewpoint. First, everything was nationalized.
Then it was realized that in this way one only hurt the living artist and
did not create sufficient resources for a national monopoly in publication. Thus royalties have been reinstated, even for the most ancient
classical texts, those already in the public domain, and those dating
from before the mediocre kind of legislation that protected writers in
Russia. Now, so it is said, the Soviets have adopted a more modern
law. In these matters, the Soviets are really hesitant, as we are in our
morality, and do not know which right should be adopted, the right of
the person, or the right over things.
6 Pirou has already made some remarks of this nature.
7 It goes without saying that we do not recommend that any dis-
mantlement of the law should take place. The principles of law that
govern the market, purchase, and sale, which are the indispensable
condition for the formation of capital, must and can subsist side by
side with new principles and more ancient ones.

  However, the moralist and the legislator must not be halted by the
so-called principles of natural law. For example, the distinction
between real law and personal law must only be considered as an

abstraction, a theoretical extract of certain of our rights. It should be allowed to subsist, but must be kept in its place.

8 Roth, 'Games', *Bull. Ethn. Queensland* (28): 23.

9 This announcement of the clan arriving is a very general custom in all eastern Australia, and is linked to the system of honour and the power inherent in the name.

10 A noteworthy fact, which leads one to surmise that wedding engagements are therefore contracted through the exchange of presents.

11 Radin, 'Winnebago Tribe', 37th *Annual Report of the Bureau of American Ethnology*, p. 320 ff.

12 See the article, 'Etiquette', in Hodge, *Handbook of American Indians*.

13 P. 326. Exceptionally, two of the chiefs invited are members of the Snake tribe.
   One can compare the speeches at a funeral festival, which are exact carbon copies of each other (tobacco). Tlingit, Swanton, 'Tlingit Myths and Texts', *Bull. of Am. Ethn.* (39): 372.

14 Rev. Taylor, *Te Ika a Maui. Old New Zealand*, p. 130, proverb 42, translates very briefly as 'give as well as take and all will be right', but the literal translation is probably as follows: As much as Maru gives, so much Maru takes, and this is good, good. (Maru is the god of war and justice.)

15 M. Bucher, *Entstehung der Volkswirtschaft*, 3rd edn, p. 73 has seen these economic phenomena, but has underestimated their importance by reducing them to hospitality.

16 *Argonauts*, p. 167 ff.; 'Primitive Economics', *Economics Journal*, March 1921. See the Preface by J.G. Frazer to Malinowski, *Argonauts*.

17 One of the extreme cases that we can cite is that of the sacrifice of dogs among the Chukchee (see Ch. 1, n. 52, p. 94). It can happen that the owners of the finest kennels massacre all their sledge teams and are forced to buy new ones.

18 See p. 44 ff.

19 See Ch. 1, p. 18; Ch. 2, n. 233, p. 131.

20 Malinowski, *Argonauts*, p. 95. Cf. Frazer, Preface to Malinowksi's book.

21 *Formes élémentaires de la vie religieuse*, p. 598, n. 2.

22 *Digesta*, XVIII, 1, *De. Contr. Emt.* 1. Paulus explains to us the great debate between the prudent Romans to settle whether *permutatio* was 'a sale'. The whole passage is interesting, even the error that the learned jurist makes in his interpretation of Homer; II, VII, 472–5:

οινίοντο does indeed mean 'to buy', but the Greek forms of money were bronze, iron, skins, even the cows and the slaves: all these had set values.

23 *Pol.*, Book 1, 1257 a. 10 ff; note the word μεταδόσιζ, ibid. 25.

24 We could just as well opt for the Arabic *sadaqa*: alms, price of the betrothed, justice, tax. Cf. p. 16.

25 *Argonauts*, p. 177.

26 It is very remarkable that in this case there is no sale, for there is no exchange of *vaygu'a*, moneys.

   The highest form of economy to which the Trobriand Islanders had raised themselves did not go as far as the use of money in exchange itself.

27 'Pure gift'.

28 P. 179.

29 The word is applied to the payment for a kind of legal prostitution for unmarried girls; cf. *Argonauts*, p. 183.

30 Cf. Ch. 2, n. 89, p. 106. The word *sagali* (cf. *hakari*) means distribution.

31 Cf. Ch. 2, n. 89, p. 107, particularly the gift of the *urigubu* to the brother-in-law: the crop produce in exchange for labour.

32 See Ch. 2, n. 86, p. 106 (*wasi*).

33 Maori. The division of labour (and the way it functions in the light of the festival between Tsimshian clans), is admirably described in a potlatch myth, Boas, 'Tsimshian Mythology', 31st *Annual Report Bur. Am. Ethn.*, pp. 274, 275; cf. p. 378. Examples of this kind could be multiplied indefinitely. These economic institutions indeed exist, even in societies infinitely less developed. See, for example, in Australia the remarkable situation of a local group that possessed a deposit of red ochre (Aisont and Horne (1924) *Savage Life in Central Australia*, London, pp. 81, 130).

34 See Ch. 2, n. 29, p. 100. The equivalence in Germanic languages of the words *token* and *Zeichen*, to designate money in general, preserves a vestige of these institutions: the sign that is the money, the sign that it carries and the pledge that it is, are one and the same thing – just as a man's signature is also what commits his responsibility.

35 See Davy, 'Foi jurée', p. 344 ff.; Davy ('Des clans aux Empires', *Eléments de Sociologie*, 1) has merely exaggerated the importance of these facts. The potlatch is useful in order to establish the hierarchy, and often does so, but it is not absolutely essential. Thus African societies, whether Nigritian or Bantu, either do not have the institution of the potlatch, or in any case have it in a not very developed

state, or have perhaps lost it – and they have all possible forms of political organization.

36 *Argonauts*, pp. 199–201; cf. p. 203.

37 Ibid, p. 199. The word 'mountain' designates in this poetry the Entrecasteaux Islands. The boat will sink under the weight of the merchandise brought back from the *kula*. Cf. another formula, p. 200, text with commentary, p. 441; cf. p. 442, where there is a remarkable play on words around 'to foam'. Cf. formula, p. 205. Cf. Ch. 2, n. 257, p. 136.

38 The area where our research would have yielded most, as well a those we have studied, is Micronesia. A system of money and contracts exists there which is extremely important, particularly in Yap and the Palaos. In Indo-China, particularly among the Mou-Khmer, in Assam, and among the Thibeto-Birmans, there are also institutions of this kind. Finally, the Berbers have developed the remarkable customs of the *thaoussa* (see Westermarck, *Marriage Ceremonies in Morocco*; see Index under the heading 'Present'). Doutté and Maunier, more competent than us, have exclusively set out to study this fact. The old Semitic law, like Bedouin customs, will also provide precious documents.

39 See the 'ritual of Beauty' in the *kula* of the Trobriand: Malinowski, p. 334 ff., p. 336: 'our partner sees us, sees that our face is beautiful, and throws to us his *vagu'a*.' Cf. Thurnwald on the use of money as an ornament, *Forschungen*, vol. 3, p. 39; cf. p. 35, the expression *Prachtbaum*, vol. 3, p. 144, lines 6, 13; p. 156, line 12 to designate a man or a woman bedecked with money. Elsewhere the chief is designated as the 'tree' (vol. 1, p. 298, line 3). Elsewhere, again, the man that is decorated exudes a perfume (vol. 1, p. 192, line 7; lines 13, 14).

40 'Fiancée market'; notion of a festival, *feria*, fair.

41 Cf. Thurnwald, ibid, vol. 3, p. 36.

42 *Argonauts*, p. 246.

43 *Salomo Inseln*, vol. 3, Table 85, n. 2.

44 Layamon's *Brut*, line 22736 ff.; *Brut*, line 9994 ff.

NAME INDEX

Subject Index

Routledge Classics
Get inside a great mind

A General Theory of Magic
Marcel Mauss

'It is enough to recall that Mauss's influence is not limited to ethnographers, none of whom could claim to have escaped it, but extends also to linguists, psychologists, historians of religion and orientalists.'
Claude Lévi-Strauss

As a study of magic in 'primitive' societies and its survival today in our thoughts and social actions, *A General Theory of Magic* represents what Claude Lévi-Strauss called the astonishing modernity of the mind of one of the century's greatest thinkers. At a period when art, magic and science appear to be crossing paths once again, *A General Theory of Magic* presents itself as a classic for our times.

Hb: 0–415–25550–3   Pb: 0–415–25396–9

Sex and Repression in Savage Society
Bronislaw Malinowski

'No writer of our times has done more than Bronislaw Malinowski to bring together in single comprehension the warm reality of human living and the cool abstractions of science.'
Robert Redfield

In *Sex and Repression in Savage Society* Malinowski, one of the founders of modern anthropology, applied his experiences on the Trobriand Islands to the study of sexuality, and the attendant issues of eroticism, obscenity, incest, oppression, power and parenthood. In so doing, he both utilized and challenged the psychoanalytical methods being popularized at the time in Europe by Freud and others. The result is a unique and brilliant book that, though revolutionary when first published, has since become a standard work on the psychology of sex.

Pb: 0–415–25554–6

For these and other classic titles from Routledge, visit
www.routledgeclassics.com